ROLLO MAY

THE DISCOVERY OF BEING

By the same author

LOVE AND WILL
THE MEANING OF ANXIETY
MAN'S SEARCH FOR HIMSELF
POWER AND INNOCENCE
THE COURAGE TO CREATE
PSYCHOLOGY AND THE HUMAN DILEMMA
FREEDOM AND DESTINY

THE DISCOVERY OF BEING

Writings in Existential Psychology

ROLLO MAY

W · W Norton & Company
NEW YORK LONDON

Published simultaneously in Canada by George J. McLeod Limited, Toronto.
Printed in the United States of America.

Grateful acknowledgment is made to Basic Books, Inc., for permission to reprint "Origins of the Existential movement in Psychology" and "Contributions of Existential Psychotherapy" from *Existence: A New Dimension in Psychology and Psychiatry,* ed. by Rollo May et al. © 1958 by Basic Books, Inc., Publishers New York.

The text of this book is composed in Avanta, with display type set in Baskerville.
Composition and manufacturing by The Haddon Craftsmen, Inc.

First Edition

Library of Congress Cataloging in Publication Data

May, Rollo.
The discovery of being.

Includes bibliographical references and index.
1. Existential psychology. 2. Existential psychotherapy. I. Title. [DNLM: 1. Existentialism. 2. Psychotherapy. BF 204.5 M467d]
BF204.5.M247 1983 150.19'2 83-4282

ISBN 0-393-01790-7

W. W. Norton & Company, Inc., 500 Fifth Avenue, New York, N.Y. 10110
W. W. Norton & Compay Ltd., 37 Great Russell Street, London WC1B 3NU

1 2 3 4 5 6 7 8 9 0

For Laura,

Companion in the search

Contents

Foreword 9

Part I: *THE PRINCIPLES*

ONE Bases of Psychotherapy 13
TWO The Case of Mrs. Hutchens 24

Part II: *THE CULTURAL BACKGROUND*

THREE Origins and Significance of Existential Psychology 37
FOUR How Existentialism and Psychoanalysis Arose Out of the Same Cultural Situation 60
FIVE Kierkegaard, Nietzsche, and Freud 67

Part III: *CONTRIBUTIONS TO THERAPY*

SIX To Be and Not to Be 91
SEVEN Anxiety and Guilt as Ontological 109

EIGHT Being in the World 117

NINE The Three Modes of World 126

TEN Of Time and History 133

ELEVEN Transcending the Immediate Situation 143

TWELVE Concerning Therapeutic Technique 151

Notes 173

Index 000

Foreword

WE IN OUR AGE are faced with a strange paradox. Never before have we had so much information in bits and pieces flooded upon us by radio and television and satellite, yet never before have we had so little inner certainty about our own being. The more objective truth increases, the more our inner certitude decreases. Our fantastically increased technical power has conferred upon us no means of controlling that power, and each forward step in technology is experienced by many as a new push toward our possible annihilation. Nietzsche was strangely prophetic when he said,

> We live in a period of atomic chaos . . . the terrible apparition . . . the Nation State . . . and the hunt for happiness will never be greater than when it must be caught between today and tomorrow, because the day after tomorrow all hunting time may have come to an end altogether.

Sensing this, and despairing of ever finding meaning in life, people these days seize on the many ways of dulling their awareness of being by apathy, by psychic numbing, or by hedonism. Others, especially young people, elect in alarming and increasing numbers to escape their own being by suicide.

No wonder people, plagued by the question of whether life has any meaning at all, flock to therapists. But therapy itself is often an expression of the fragmentation of our age rather than an enterprise for overcoming it. Often these persons, seeking release from their feelings of emptiness on the couch or in the client's chair, surrender their being to the therapist —which can only lead to a submerged despair, a burrowing resentment that will later burst out in self-destructiveness. For history proclaims again and again that sooner or later the individual's need to be free will assert itself.

I believe it is by discovering and affirming the being in ourselves that some inner certainty will become possible. In contrast to the psychologies that conclude with theories about conditioning, mechanisms of behavior, and instinctual drives, I maintain that we must go below these theories and discover the person, *the being to whom these things happen.*

True, we all seem in our culture to be hesitant to talk of being. Is it too revealing, too intimate, too profound? In covering up being we lose just those things we most cherish in life. For the sense of being is bound up with the questions that are deepest and most fundamental—questions of love, death, anxiety, caring.

The writings in this book have grown out of my passion to find the being in my fellow persons and myself. This always involves the search for our values and purposes. In the experience of normal anxiety, for example, if the person did not have anxiety, he or she would also not have freedom. Anxiety demonstrates that values, no matter how beclouded, do exist in the person. Without values there would be only barren despair.

As we face the severest threat in history to human survival, I find the possibilities of being made more prominent by their contrast with our possible annihilation. The individual human is still the creature who can wonder, who can be enchanted by a sonata, who can place symbols together to make poetry to gladden our hearts, who can view a sunrise with a sense of majesty and awe.

All of these are characteristic of being, and they set the challenge for the pages that follow.

PART I

THE PRINCIPLES

ONE

Bases of Psychotherapy

THOUGH THE existential approach had been the most prominent in European psychiatry and psychoanalysis for two decades, it was practically unknown in America until 1960. Since then, some of us have been worried that it might become *too* popular in some quarters, particularly in national magazines. But we have been comforted by a saying of Nietzsche's: "The first adherents of a movement are no argument against it."

In the United States there is, paradoxically, both an affinity and an aversion to existential therapy. On the one hand, this approach has a deep underlying affinity for our American character and thought. It is very close, for example, to William James's emphases on the immediacy of experience, the unity of thought and action, and the importance of decision and commitment. On the other hand, there is among some psychologists and psychoanalysts in this country a great deal of hostility toward and outright anger against this approach. I shall later go into reasons for this paradox.

I wish in this volume, rather, to *be* existential and to speak directly from my own experience as a person and as a practicing psychoanalytic psychotherapist. When I was working on *The Meaning of Anxiety,* I spent a year and a half in bed in a tuberculosis sanatorium. I had a great deal of time to ponder the meaning of anxiety—and plenty of firsthand data on myself and my anxious fellow patients. In the course of this time I studied the two books written on anxiety up till our day: one by Freud, *The Problem of Anxiety,* and the other by Kierkegaard, *The Concept of Anxiety.* I valued highly Freud's formulations—for example, his first theory, that anxiety is the reemergence of repressed libido, and his second, that anxiety is the ego's reaction to the threat of the loss of the loved object. But these still were theories. Kierkegaard, on the other hand, described anxiety as the struggle of the living being against nonbeing which I could immediately experience in my struggle with death or the prospect of being a lifelong invalid. Kierkegaard went on to point out that the real terror in anxiety is not death as such but the fact that each of us within himself is on both sides of the fight, that "anxiety is a desire for what one dreads," as he put it; thus like an "alien power it lays hold of an individual, and yet one cannot tear one's self away."

What powerfully struck me then was that Kierkegaard was writing about *exactly what my fellow patients and I were going through.* Freud was not; he was writing on a different level, giving formulations of the psychic mechanisms by which anxiety comes about. Kierkegaard was portraying what is immediately experienced by human beings in crisis—the crisis specifically of life against death which was completely real to us patients, but a crisis which I believe is not in its essential form different from the various crises of people who come for therapy, or the crises all of us experience in much more minute form a dozen times a day even though we push the ultimate prospect of death far from our minds. Freud was writing on the technical level, where his genius was supreme; perhaps more than any man up to his time, he *knew about* anxiety. Kierkegaard, a genius of a different order, was writing on the existential, ontological level; he *knew anxiety.*

This is not a value dichotomy; obviously both are necessary. Our real problem, rather, is given us by our cultural-historical situation. We in the Western world are the heirs of four centuries of technical achievement in power over nature, and now over ourselves; this is our greatness and, at the same time, it is also our greatest peril. We are not in danger of repressing the technical emphasis (of which Freud's tremendous popularity in this country was proof, if any were necessary). But rather we repress the opposite. If I may use terms which I shall be discussing more fully presently, we repress the *sense of being,* the ontological sense. One consequence of this repression of the sense of being is that modern man's image of himself, his experience of himself as a responsible individual, his experience of his own humanity have likewise disintegrated.

The existential approach does not have the aim of ruling out the technical discoveries of Freud or those from any other branch of psychology or science. It does, however, seek to place these discoveries on a new basis, a new understanding or rediscovery of the nature and image of the human being.

I make no apologies in admitting that I take very seriously the dehumanizing dangers in our tendency in modern science to make man over into the image of the machine, into the image of the techniques by which we study him. This tendency is not the fault of any "dangerous" person or "vicious" schools. It is rather a crisis brought upon us by our particular historical predicament. Karl Jaspers, psychiatrist and existentialist philosopher, held that we in the Western world are actually in process of losing self-consciousness and that we may be in the last age of historical man. William Whyte in his *Organization Man* cautioned that modern man's enemies may turn out to be a "mild-looking group of therapists, who . . . would be doing what they did to help you." He was referring to the tendency to use the social sciences in support of the social ethic of our historical period; thus the process of helping people may actually make them conformist and tend toward the destruction of individuality. This tendency, I believe, increases radically with the spread of behavior modification, a form of psychotherapy based on an outspoken denial of any need for a theory of man at all beyond the therapist's assumption that whatever goals he and his group have chosen are obviously the best for all possible human beings. We cannot brush aside the cautions of such men as Jaspers and Whyte as unintelligent or antiscientific. To try to do so would make *us* the obscurants.

Many psychologists share my sentiments but cavil at the terms "being" and "nonbeing", concluding that the existential approach in psychology is hopelessly vague and muddled. But I would hold that *without* some concepts of "being" and "nonbeing," we cannot even understand our most commonly used psychological mechanisms. Take, for example, *repression* and *transference.* The usual discussions of these terms hang in mid-air, without convincingness or psychological reality precisely because we have lacked an underlying structure on which to base them. The term *repression* obviously refers to a phenomenon we observe all the time, a dynamism which Freud clearly described in many forms. We generally explain the mechanism by saying that the child represses into unconsciousness certain impulses, such as sex and hostility, because the culture in the form of parental figures disapproves, and the child must protect his own security with these persons. But this culture which assumedly disapproves is made up of the very same people who do the repressing. Is it not an illusion, therefore, and much too simple, to speak of the culture over against the individual in such fashion and make it our whipping boy? Furthermore, where did we get the ideas that child or adult are so much concerned with security and libidinal satisfactions? Are these not a carry-over from our work with the *neurotic, anxious* child and adult?

Certainly the neurotic, anxious child is compulsively co cerned with security, for example; and certainly the neuroti adult, and we who study him, read our later formulations back into the unsuspecting mind of the child. But is not the normal child just as truly interested in moving out into the world, exploring, following his curiosity and sense of adventure—going out "to learn to shiver and to shake," as the nursery rhyme puts it? And if you block these needs of the child, you get a traumatic reaction from him just as you do when you take away his security. I, for one, believe we vastly overemphasize the human being's concern with security and survival satisfactions because they so neatly fit our cause-and-effect way of thinking. I believe Nietzsche and Kierkegaard were more accurate when they described man as the organism who makes certain values—prestige, power, tenderness—more important than pleasure and even more important than survival itself.

My thesis here is that we can understand repression, for example, only on the deeper level of the meaning of the human being's potentialities. In this respect, "being" is to be defined as the individual's "pattern of potentialities." These potentialities will be partly shared with other persons but will in every case form a unique pattern in each individual. We must ask the questions: What is this person's relation to his own potentialities? What goes on that he chooses or is forced to choose to block off from his awareness something which he knows, and on another level *knows that he knows?* In my work in psychotherapy there appears more and more evidence that anxiety in our day arises not so much out of fear of lack of libidinal satisfactions or security, but rather out of the patient's fear of his own powers, and the conflicts that arise from that fear. This may be the particular "neurotic personality of our time"—the neurotic pattern of contemporary "outer-directed" organizational man.

The "unconscious," then, is not to be thought of as a reservoir of impulses, thoughts, wishes which are culturally unacptable. I define it rather as *those potentialities for knowing nd experiencing which the individual cannot or will not actualize.* On this level we shall find that the simple mechanism of repression is infinitely less simple than it looks; that it involves a complex struggle of the individual's *being* against the possibility of *nonbeing;* that it cannot be adequately comprehended in "ego" and "not-ego" terms, or even "self" and "not-self"; and that it inescapably raises the question of the human being's margin of freedom with respect to his potentialities, a margin in which resides his responsibility for himself which even the therapist cannot take away.

Another concept from classical analysis besides repression bears comment here. I refer to transference, the relationship between the two people, patient and therapist, in the consulting room. The concept and description of transference was one of Freud's great contributions, both in his own judgment and in that of many of the rest of us. There are vast implications for therapy in the phenomenon that the patient brings into the consulting room his previous or present relationships with father, mother, lover, child, and proceeds to perceive us as those creatures and to build his world with us in the same way. Transference, like other concepts of Freud, vastly enlarges the sphere and influence of personality; we live in others and they in us. Note Freud's idea that in every act of sexual intercourse four persons are present—one's self and one's lover, plus one's two parents. I have always personally taken an ambivalent attitude toward this idea, believing as I do that the act of love at least deserves some privacy. But the deeper implications are the fateful interweaving of the human web; one's ancestors, like Hamlet's father, are always coming on to the edge of the stage with various ghostly challenges and imprecations. This emphasis of Freud's on how deeply we are bound each to each again cuts through many of modern man's illusions about love and interpersonal relations.

But the concept of transference presents us with unending

difficulties if we take it by itself, i.e., without a nor
ship which is grounded in the nature of man as su
first place, transference can be a handy and ever-usefu
for the therapist, as Thomas Szasz puts it; the therap
hide behind it to protect himself from the anxiety of d
encounter. Second, the concept of transference can undermi
the whole experience and sense of reality in therapy; the two
persons in the consulting room become "shadows," and everyone else in the world does too. It can erode the patient's sense of responsibility, and can rob the therapy of much of the dynamic for the patient's change.

What has been lacking is a concept of *encounter,* within which, and only within which, transference has genuine meaning. *Transference is to be understood as the distortion of encounter.* Since there was no norm of human encounter in psychoanalysis and no adequate place for the *I*-thou relationship, there was bound to be an oversimplifying and watering down of love relationships. Freud greatly deepened our understanding of the multifarious, powerful and ubiquitous forms in which erotic drives express themselves. But eros (instead of coming back into its own, as Freud fondly hoped) now occillates between being an absurd chemistry that demands outlet and a relatively unimportant pastime for male and female of an evening when they get bored watching TV.

Also, we had no norm of *agape* (the form of selfless love, concern for the other person's welfare) in its own right. Agape cannot be understood as derivative, or what is left over when you analyze out exploitative, cannibalistic tendencies. Agape is not a sublimation of eros but a transcending of it in enduring tenderness, lasting concern for the other. And it is precisely this transcendence which gives eros itself fuller and more enduring meaning.

The existential approach helps us in asking the question: How is it possible that one being relates to another? What is the nature of human beings that two persons can communi-

...sp each other as beings, have genuine concern with ...e and fulfillment of the other, and experience some ... trust? The answer to these questions will tell us of ... transference is a distortion.

...s I sit now in relationship with my patient, I am assuming ...hat this man, let us say, like all existing beings, needs to reach out from his own centeredness to participate with other persons. Before he ever made the tentative and oft-postponed steps to phone me for an appointment, he was already participating in imagination in some relationship with me. He sat nervously smoking in my waiting room; he now looks at me with mingled suspicion and hope, an effort toward openness fighting in him against the life-old tendency to withdraw behind a stockade and hold me out. This struggle is understandable, for *participating always involves risk.* If one goes out too far, one will lose one's identity. But if he is so afraid of losing his own conflicted center—which at least has made possible some partial integration and meaning in his experience—that he refuses to go out at all but holds back in rigidity and lives in narrowed and shrunken world space, his growth and development are blocked. This is what Freud meant when he spoke of repression and inhibition. Inhibition is the relation to the world of the being who has the possibility to go out but is too threatened to do so; and his fear that he will lose too much may, of course, be the case. Patients will say, "If I love somebody, it's as though all of me will flow out like water out of a river, and there'll be nothing left." I think this is a very accurate statement of *transference.* That is, if one's love is something that does not belong there of its own right, then obviously it will be emptied. The whole matter is one of economic balance, as Freud put it.

But in our day of conformism and the outer-directed man, the most prevalent neurotic pattern takes the opposite form—namely, going out too far, dispersing one's self in participation and identification with others until one's own being is emptied.

This is the psycho-cultural phenomenon of the organization man. It is one reason that castration is no longer the dominant fear of men or women in our day, but ostracism. Patient after patient I've seen (especially those from Madison Avenue) chooses to be castrated—that is, to give up his power—in order not to be ostracized. The real threat is not to be accepted, to be thrown out of the group, to be left solitary and alone. In this overparticipation, one's own consistency becomes inconsistent because it fits someone else. One's own meaning becomes meaningless because it is borrowed from somebody else's meaning.

Speaking now more concretely of the concept of encounter, I mean it to refer to the fact that in the therapeutic hour a total relationship is going on between two people which includes a number of different levels. One level is that of real persons: I am glad to see my patient (varying on different days depending chiefly on the amount of sleep I have had the night before). Our seeing each other allays the physical loneliness to which all human beings are heir. Another level is that of *friends:* we trust—for we have seen a lot of each other—that the other has some genuine concern for listening and understanding. Another level is that of *esteem,* or *agape,* the capacity which inheres in *Mitwelt** for self-transcending concern for another's welfare. Another level will be frankly *erotic.* When I was doing supervision with her some years ago, Clara Thompson once said to me something I've often pondered, that if one person in the therapeutic relationship feels active erotic attraction, the other will too. Erotic feelings of his own need to be frankly faced by the therapist; otherwise he will, at least in fantasy, act out his own needs with the patient. But more importantly, unless the therapist accepts the erotic as one of the ways of

*This German word, *Mitwelt,* means literally the "with world," the world of interpersonal relations. The word is explained fully, along with the two similar German words, *Umwelt,* the "around-world" or environment, and *Eigenwelt,* the world within oneself, in Chapter 9.

communication, he will not listen for what he should hear from the patient and he will lose one of the most dynamic resources for change in therapy.

Now this total encounter, which can be our most useful medium of understanding the patient as well as our most efficacious instrument for helping him open himself to the possibility of change, seems to me to have the resonant character of two musical instruments. If you pluck a violin string, the corresponding strings in another violin in the room will resonate with corresponding movement of their own. This is an analogy, of course: what goes on in human beings includes that, but is much more complex. Encounter in human beings is always to a greater or lesser extent *anxiety-creating* as well as *joy-creating.* I think these effects arise out of the fact that genuine encounter with another person always shakes our self-world relationship: our comfortable temporary security of the moment before is thrown into question, we are opened, made tentative for an instant—shall we risk ourselves, take the chance to be enriched by this new relationship (and even if it is a friend or loved one of long standing, this particular moment of relationship is still new) or shall we brace ourselves, throw up a stockade, block out the other person and miss the nuances of his perceptions, feelings, intentions? Encounter is always a potentially creative experience; it normally should ensue in the expanding of consciousness, the enrichment of the self. (I do not speak here of *quantity*—obviously a brief meeting may affect us only slightly; indeed, I do not refer to quantities at all, but to a *quality* of experience.) In genuine encounter both persons are changed, however minutely. C. G. Jung has pointed out rightly that in effective therapy a change occurs in *both* the therapist and the patient; unless the therapist is open to change the patient will not be either.

The phenomenon of encounter very much needs to be studied, for it seems clear that much more is going on than almost any of us has realized. I propose the hypothesis that in therapy, granted adequate clarification of the therapist, *it is not possible for one person to have a feeling without the other having it to some degree also.* I know there will be many exceptions to this, but I want to offer the hypothesis to ponder and work on. One corollary of my hypothesis is that in *Mitwelt* there is necessarily some resonance, and that the reason we don't feel it, when we don't, is some blocking on our part. Frieda Fromm-Reichman often said that her best instrument for telling what the patient feels—e.g., anxiety or fear or love or anger that he, the patient, dare not express—*is what she feels at that moment within herself.* This use of oneself as the instrument, of course, requires a tremendous self-discipline on the part of the therapist. I don't mean at all to open the door simply to telling the patient what you, the therapist, feel. Your feelings may be neurotic in all sorts of ways, and the patient has enough problems without being further burdened with yours. I mean rather that the self-discipline, the self-purification if you will, the bracketing of one's own distortions and neurotic tendencies to the extent a therapist is able, seems to me to result in his being in greater or lesser degree able to experience encounter as a way of participating in the feelings and the world of the patient. All this needs to be studied and I believe can be studied in many more ways than we have realized. As I have said, there is something going on in one human being relating to another, something inhering in *Mitwelt,* that is infinitely more complex, subtle, rich, and powerful than we have realized.

The chief reason this hasn't been studied, it seems to me, is that we have had no concept of encounter, for it was covered up by Freud's concept of transference. As one consequence, we have had all kinds of studies of transference, which tell us everything except what really goes on between two human beings. We are justified in looking to phenomenology for help in arriving at a concept which will enable us *to perceive encounter itself when so far we have only perceived its distortion, transference.* It is especially important that we not yield to the tendency to avoid and dilute encounter by making it a derivative of transference or countertransference.

TWO

The Case of Mrs. Hutchens

AS A PRACTICING therapist and teacher of therapists, I have been struck by how often our concern with trying to understand the patient in terms of the mechanisms by which his behavior takes place blocks our understanding of what he really is experiencing. A patient, Mrs. Hutchens (about whom I shall center some of my remarks), comes into my office for the first time, a suburban woman in her middle thirties who tries to keep her expression poised and sophisticated. But no one could fail to see in her eyes something of the terror of a frightened animal or a lost child. I know, from what her neurological specialists have already told me, that her presenting problem is hysterical tenseness of the larynx, as a result of which she can talk only with a perpetual hoarseness. I have been given the hypothesis from her Rorschach that she has felt all her life that "If I say what I really feel, I'll be rejected; under these conditions it is better not to talk at all." During this first hour, also, I get some hints of the genetic *why* of her problem as she tells me of her authoritarian relation with her mother and grandmother, and how she learned to guard firmly against telling any secrets at all. But if as I sit here I am chiefly thinking of these why's and how's concerning the way the problem came about, I will grasp everything except the most important thing of all (indeed the only real source of data I have), namely, this person now existing, becoming, emerging—this experiencing human being immediately in the room with me.

There are in this country several undertakings to systematize psychoanalytic theory in terms of forces, dynamisms, and energies. The approach I propose is the exact opposite of this. I hold that our science must be relevant to the distinctive characteristics of what we seek to study, in this case the human being. I do not deny dynamisms and forces—that would be nonsense—but I hold that they have meaning only in the context of the existing, living person, that is to say, in the *ontological* context.

I propose, thus, that we take the one real datum we have in the therapeutic situation, namely, the *existing person* sitting in a consulting room with a therapist. (The term "existing person" is my equivalent of the German *Dasein,* literally the being who is there.) Note that I do not say simply "individual" or "person"; if you take individuals as units in a group for the purposes of statistical prediction—certainly a legitimate use of psychological science—you are exactly *defining out of the picture* the characteristics which make this individual an existing person. Or when you take him or her as a composite of drives and deterministic forces, you have defined for study everything except *the one to whom these experiences happen,* everything except the existing person himself. Therapy is one activity in which we cannot escape the necessity of taking the subject as an existing person.

Let us, therefore, ask: What are the essential characteristics which constitute this patient as an existing person in the consulting room? I wish to propose six characteristics which I shall call principles,[1] which I find in my work as a psychotherapist. Though these principles are the product of a good deal of thought and experience with many cases, I shall illustrate them with episodes from the case of Mrs. Hutchens.

First, Mrs. Hutchens like, every existing person, *is centered in herself,* and an attack on this center is an attack on her existence itself. This is a characteristic which we share with all living beings; it is self-evident in animals and plants. I never cease to marvel how, whenever we cut the top off a pine tree on our farm in New Hampshire, the tree sends up a new branch from heaven knows where to become a new center. But this principle has a particular relevance to human beings and gives a basis for the understanding of sickness and health, neurosis and mental health. Neurosis is not to be seen as a deviation from our particular theories of what a person should be. *Is not neurosis, rather, precisely the method the individual uses to preserve his own center, his own existence?* His symptoms are ways of shrinking the range of his world (so graphically shown in Mrs. Hutchens's inability to let herself talk) in order that the centeredness of his existence may be protected from threat; a way of blocking off aspects of the environment that he may then be adequate to the remainder. Mrs. Hutchens had gone to another therapist for half a dozen sessions a month before she came to me. He told her, in an apparently ill-advised effort to reassure her, that she was too proper, too controlled. She reacted with great upset and immediately broke off the treatment. Now technically he was entirely correct; existentially he was entirely wrong. What he did not see, in my judgment, was that this very properness, this overcontrol, far from being things Mrs. Hutchens wanted to get over, were part of her desperate attempt to preserve what precarious center she had. As though she were saying, "If I opened up, if I communicated, I would lose what little space in life I have."

We see here, incidentally, how inadequate is the definition of neurosis as a failure of adjustment. *An adjustment is exactly what neurosis is; and that is just its trouble.* It is a necessary adjustment by which centeredness can be preserved; a way of accepting *nonbeing* in order that some little *being* may be preserved. And in most cases it is a boon when this adjustment breaks down.

This is the only thing we can assume about Mrs. Hutchens, or about any patient, when she comes in: that she, like all living beings, requires centeredness, and that this has broken down. At a cost of considerable turmoil she has taken steps—that is, come for help. Thus, our second principle is: *every existing person has the character of self-affirmation, the need to preserve his centeredness.* The particular name we give this self-affirmation in human beings is "courage." Paul Tillich's writing on the "courage to be" is very cogent and fertile for psychotherapy at this point. He insists that in human being is never given automatically but depends upon the individual's courage, and without courage one loses being. *This makes courage itself a necessary ontological corollary.* By this token, I as a therapist place great importance upon expressions of the patients which have to do with willing, decisions, choice. I never let little remarks the patient may make such as "maybe I can," "perhaps I can try," and so on slip by without my making sure he knows I have heard him. It is only a half truth that the will is the product of the wish; I emphasize rather the truth that the wish can never come out in its real power except with will.

Now as Mrs. Hutchens talks hoarsely, she looks at me with an expression of mingled fear and hope. Obviously a relation exists between us not only here but already in anticipation in the waiting room and ever since she thought of coming. She is struggling with the possibility of participating with me. Our third principle is, thus: *all existing persons have the need and possibility of going out from their centeredness to participate in other beings.* This always involves risk. If the organism goes out too far, it loses its own centeredness—its identity—a phenomenon which can easily be seen in the biological world. The gypsy moth, for example, increases phenomenally for several years, eating the leaves off trees at a fantastic rate, eventually eating itself out of its own food and dying out.

But if the neurotic is so afraid of loss of his own center, conflicted though it be, that he refuses to go out but holds back in rigidity and lives in narrowed reactions and shrunken world space, his growth and development are blocked, as we noted in Chapter 1. This is the pattern in neurotic repressions and inhibitions, the common neurotic forms in Freud's day. But it may well be in our day of conformism and the outer-directed man, that the most common neurotic pattern takes the opposite form—namely, the dispersing of oneself in participation and identification with others until one's own being is emptied. Like the gypsy moth, we destroy our own being. At this point we see the rightful emphasis of Martin Buber in one sense and Harry Stack Sullivan in another, that the human being cannot be understood as a self if participation with other selves is omitted. Indeed, if we are successful in our search for these ontological principles of the existing person, it should be true that the omission of any one of the six would mean we do not then have a human being.

Our fourth principle is: *the subjective side of centeredness is awareness.* The paleontologist Pierre Teilhard de Chardin described brilliantly how this awareness is present in ascending degrees in all forms of life from amoeba to man. It is certainly present in animals. Howard Liddell has pointed out how the seal in its natural habitat lifts its head every ten seconds even during sleep to survey the horizon lest an Eskimo hunter with poised bow and arrow sneak up on it. This awareness of threats to being in animals Liddell calls *vigilance,* and he identifies it as the primitive, simple counterpart in animals of what in human beings becomes anxiety.

Our first four characteristic principles are shared by our existing person with all living beings; they are biological levels in which human beings participate. The fifth principle refers now to a distinctively human characteristic, self-consciousness. *The uniquely human form of awareness is self-consciousness.*

We do not identify awareness and consciousness. We associate awareness, as Liddell indicates above, with vigilance. This is supported by the derivation of the term—it comes from the Anglo-Saxon *gewaer, waer,* meaning knowledge of external dangers and threats. Its cognates are *beware* and *wary.* Awareness certainly is what is going on in an individual's neurotic reaction to threat, in Mrs. Hutchens's experience in the first hours, for example, that I am also a threat to her. Consciousness, in contrast, we define as not simply my awareness of threat from the world, but *my capacity to know myself as the one being threatened,* my experience of myself as the subject who has a world. Consciousness, as Kurt Goldstein puts it, is man's capacity to transcend the immediate concrete situation, to live in terms of the possible; and it underlies the human capacity to use abstractions and universals, to have language and symbols. This capacity for consciousness underlies the wide range of possibility which man has in relating to his world, and it constitutes the foundation of psychological freedom. Thus human freedom has its ontological base and I believe must be assumed in all psychotherapy.

In his book *The Phenomenon of Man,* Pierre Teilhard de Chardin, as we have mentioned, describes awareness in all forms of evolutionary life. But in man, a new function arises —namely, this self-consciousness. Teilhard de Chardin undertakes to demonstrate something I have always believed, that when a new function emerges the whole previous pattern, the total Gestalt of the organism, changes. Thereafter the organism can be understood only in terms of the new function. That is to say, it is only a half truth to hold that the organism is to be understood in terms of the simpler elements below it on the evolutionary scale; it is just as true that every new function forms a new complexity which conditions all the simpler elements in the organism. *In this sense, the simple can be understood only in terms of the more complex.*

This is what self-consciousness does in man. All the simpler biological functions must now be understood in terms of the new function. No one would, of course, deny for a moment the old functions, nor anything in biology which man shares with less complex organisms. Take sexuality, for example, which we obviously share with all mammals. But given self-consciousness, sex becomes a new Gestalt as is demonstrated in therapy all the time. Sexual impulses are now conditioned by the *person* of the partner; what we think of the other male or female, in reality or fantasy or even repressed fantasy, can never be ruled out. The fact that the subjective person of the other to whom we relate sexually makes least difference in *neurotic* sexuality, say in patterns of compulsive sex or prostitution, only proves the point the more firmly; for such requires precisely the blocking off, the checking out, the distorting of self-consciousness. Thus when we talk of sexuality in terms of sexual *objects,* as Kinsey did, we may garner interesting and useful statistics; but we simply are not talking about human sexuality.

Nothing in what I am saying here should be taken as antibiological in the slightest; on the contrary, I think it is only from this approach that we *can* understand human biology without distorting it. As Kierkegaard aptly put it, "The natural law is as valid as ever." I argue only against the uncritical acceptance of the assumption that the organism is to be understood solely in terms of those elements below it on the evolutionary scale, an assumption which has led us to overlook the self-evident truth that what makes a horse a horse is not the elements it shares with the organisms below it but what constitutes distinctively "horse." Now *what we are dealing with in neurosis are those characteristics and functions which are distinctively human.* It is these that have gone awry in our disturbed patients. The condition for these functions is self-consciousness —which accounts for what Freud rightly discovered, that the neurotic pattern is characterized by repression and blocking off of consciousness.

It is the task of the therapist, therefore, not only to help the patient become aware, but even more significantly to help him to *transmute this awareness into consciousness.* Awareness is his knowing that something is threatening from outside in his world—a condition which may, as in paranoids and their neurotic equivalents, be correlated with a good deal of acting-out behavior. But self-consciousness puts this awareness on a quite different level; it is the patient's seeing that *he is the one who is threatened,* that he is the being who stands in this world which threatens, he is the subject who *has* a world. And this gives him the possibility of *in-sight,* of "inward sight," of seeing the world and its problems in relation to himself. And thus it gives him the possibility of doing something about the problems.

To come back to our too-long silent patient, after about twenty-five hours of therapy Mrs. Hutchens had the following dream. She was searching room by room for a baby in an unfinished house at an airport. She thought the baby belonged to someone else, but the other person might let her borrow it. Now it seemed that she had put the baby in a pocket of her robe (or her mother's robe) and she was seized with anxiety that it would be smothered. Much to her joy, she found that the baby was still alive. Then she had a strange thought, "Shall I kill it?"

The house was at the airport where she at about the age of twenty had learned to fly solo, a very important act of self-affirmation and independence from her parents. The baby was associated with her youngest son, whom she regularly identified with herself. Permit me to omit the ample associative evidence that convinced both her and me that the baby stood for herself. The dream is an expression of the emergence and growth of self-consciousness, a consciousness she is not sure is hers yet, and a consciousness which she considers killing in the dream.

About six years before her therapy, Mrs. Hutchens had left the religious faith of her parents, to which she had had a very authoritarian relation. She had then joined a church of her own belief. But she had never dared tell her parents of this. Instead, when they came to visit, she attended their church in great tension lest one of her children let the secret out. After about thirty-five sessions, when she was considering writing her parents to tell them of this change of faith, she had over a period of two weeks spells of partially fainting in my office. She would become suddenly weak, her face would go white, she would feel empty and "like water inside," and would have to lie down for a few moments on the couch. In retrospect she called these spells "grasping for oblivion."

She then wrote her parents informing them once and for all of her change in faith and assuring them it would do no good to try to dominate her. In the following session she asked in considerable anxiety whether I thought she would go psychotic. I responded that whereas any one of us might at some time have such an episode, I saw no more reason why she should than any of the rest of us. And I asked whether her fear of going psychotic was not rather anxiety coming out of her standing against her parents, as though genuinely being herself she felt to be tantamount to going crazy. I have, it may be remarked, several times noted this anxiety at being oneself experienced by the patient as tantamount to psychosis. This is not surprising, for consciousness of one's own desires and affirming them involves accepting one's originality and uniqueness, and it implies that one must be prepared to be isolated not only from those parental figures upon whom one has been dependent, but at that instant to stand alone in the entire psychic universe as well.

We see the profound conflicts of the emergence of self-consciousness in three vivid ways in Mrs. Hutchens, whose chief symptom, interestingly enough, was the denial of that uniquely human capacity based on consciousness—namely, talking: (1) the temptation to kill the baby; (2) the grasping at oblivion by fainting, as though she were saying, "If only I did not have to be conscious, I would escape this terrible problem of telling my parents"; and (3) the psychosis anxiety.

We now come to the sixth and last ontological characteristic, *anxiety. Anxiety is the state of the human being in the struggle against what would destroy his being.* It is, in Tillich's phrase, the state of a being in conflict with nonbeing, a conflict which Freud mythologically pictured in his powerful and important symbol of the death instinct. One wing of this struggle will always be against something outside oneself. But even more portentous and significant for psychotherapy is the inner side of the battle, which we saw in Mrs. Hutchens—namely, the conflict within the person as she confronts the choice of whether and how far she will stand against her own being, her own potentialities.

From an existential viewpoint we take very seriously this temptation to kill the baby, or kill her own consciousness, as expressed in these forms by Mrs. Hutchens. We neither water it down by calling it "neurotic" and the product merely of sickness, nor slough over it by reassuring her, "O.K., but you don't need to do it." If we did these, we would be helping her adjust at the price of surrendering a portion of her existence—that is, her opportunity for fuller independence. The self-confrontation which is involved in the acceptance of self-consciousness is anything but simple: it involves, to identify some of the elements, accepting the hatred of the past, her mother's of her and hers of her mother; accepting her present motives of hatred and destruction; cutting through rationalizations and illusions about her behavior and motives, and the acceptance of the responsibility and aloneness which this implies; the giving up of childhood omnipotence, and acceptance of the fact that though she can never have absolute certainty of choices, she must choose anyway. But all of these specific points, easy enough to understand in themselves, must be seen in the light of the fact that *consciousness itself implies always the possibility of turning against oneself, denying oneself.* The tragic nature of human existence inheres in the fact that consciousness itself involves the possibility and temptation at every instant of killing itself. Dostoevski and our other existential forebears were not indulging in poetic hyperbole or expressing the aftereffects of immoderate vodka when they wrote of the agonizing burden of freedom.

The fact that existential psychotherapy places emphasis on these tragic aspects of life does not at all imply it is pessimistic. Quite the contrary. The confronting of genuine tragedy is a highly cathartic experience psychically, as Aristotle and others through history have reminded us. Tragedy is inseparably connected with the human being's dignity and grandeur, and is the accompaniment, as illustrated in the dramas of Oedipus and Orestes and *Hamlet* and *Macbeth,* of the person's moments of greatest insight.

PART II

THE CULTURAL BACKGROUND

THREE

Origins and Significance of Existential Psychology

IN RECENT YEARS there has been a growing awareness on the part of some psychiatrists and psychologists that serious gaps exist in our way of understanding human beings. These gaps may well seem most compelling to psychotherapists, confronted as they are in clinic and consulting room with the sheer reality of persons in crisis whose anxiety will not be quieted by theoretical formulas. But the lacunae likewise present seemingly unsurmountable difficulties in scientific research. Thus many psychiatrists and psychologists in Europe and others in this country have been asking themselves disquieting questions, and others are aware of gnawing doubts which arise from the same half-suppressed and unasked questions.

Can we be sure, one such question goes, that we are seeing the patient as he really is, knowing him in his own reality; or are we seeing merely a projection of our own theories *about* him? Every psychotherapist, to be sure, has his knowledge of patterns and mechanisms of behavior and has at his fingertips the system of concepts developed by his particular school. Such a conceptual system is entirely necessary if we are to observe scientifically. But the crucial question is always the bridge between the system and the patient—how can we be certain that our system, admirable and beautifully wrought as it may be in principle, has anything whatever to do with this specific Mr. Jones, a living, immediate reality sitting opposite us in the consulting room? May not just this particular person require another system, another quite different frame of reference? And does not this patient, or any person for that matter, evade our investigations, slip through our scientific fingers like sea foam, precisely to the extent that we rely on the logical consistency of our own system?

Another such gnawing question is: How can we know whether we are seeing the patient in his real world, the world in which he "lives and moves and has his being," and which is for him unique, concrete, and different from our general theories of culture? In all probability we have never participated in his world and do not know it directly. Yet we must know it and to some extent must be able to exist in it if we are to have any chance of knowing the patient.

Such questions were the motivations of psychiatrists and psychologists in Europe, who later comprised the *Daseinsanalyse,* or existential-analytic, movement. The "existential research orientation in psychiatry," writes Ludwig Binswanger, its chief spokesman, "arose from dissatisfaction with the prevailing efforts to gain scientific understanding in psychiatry. . . . Psychology and psychotherapy as sciences are admittedly concerned with 'man,' but not at all primarily with mentally *ill* man, but with *man as such.* The new understanding of man, which we owe to Heidegger's analysis of existence, has its basis in the new conception that man is no longer understood in terms of some theory—be it a mechanistic, a biologic or a psychological one."[1]

WHAT CALLED FORTH THIS DEVELOPMENT?

Before turning to what this new conception of man is, let us note that this approach sprang up spontaneously in different parts of Europe and among different schools, and has a diverse body of researchers and creative thinkers. There were Eugene Minkowski in Paris, Erwin Straus in Germany and later in this country, V. E. von Gebsattel in Germany, who represented chiefly the first, or phenomenological, stage of this movement. There were Ludwig Binswanger, A. Storch, M. Boss, G. Bally, Roland Kuhn in Switzerland, J. H. Van Den Berg and F. J. Buytendijk in Holland, and so on, representing more specifically the second, or existential, stage. These facts—namely, that the movement emerged spontaneously, without these men in some cases knowing about the remarkably similar work of their colleagues, and that, rather than being the brainchild of one leader, it owes its creation to diverse psychiatrists and psychologists—testify that it must answer a widespread need in our times in the fields of psychiatry and psychology. Von Gebsattel, Boss, and Bally are Freudian analysts; Binswanger, though in Switzerland, became a member of the Vienna Psychoanalytic Society at Freud's recommendation when the Zurich group split off from the International. Some of the existential therapists had also been under Jungian influence.

These thoroughly experienced men became disquieted over the fact that, although they were effecting cures by the techniques they had learned, they could not, so long as they confined themselves to Freudian and Jungian assumptions, arrive at any clear understanding of why these cures did or did not occur or what actually was happening in the patients' existence. They refused the usual methods among therapists of quieting such inner doubts—namely, of turning one's attention with redoubled efforts to perfecting the intricacies of one's own conceptual system. Another tendency among psychotherapists, when anxious or assailed by doubts as to what they are doing, is to become preoccupied with technique. Perhaps the most handy anxiety-reducing agent is to abstract oneself from the issues by assuming a wholly technical emphasis. These men resisted this temptation. They likewise were unwilling to postulate unverifiable agents, such as "libido," or "censor," as Ludwig Lefebre points out,[2] or the various processes lumped under "transference," to explain what was going on. And they had particularly strong doubts about using the theory of the unconscious as a carte blanche on which almost any explanation could be written. They were aware, as Straus puts it, that the "unconscious ideas of the patient are more often than not the conscious theories of the therapist."

It was not with specific techniques of therapy that these psychiatrists and psychologists took issue. They recognized for example, that psychoanalysis is valid for certain types of cases, and some of them, bona fide members of the Freudian movement, employed it themselves. But they all had grave doubts about its theory of man. And they believed these difficulties and limitations in the concept of man not only seriously blocked research but would in the long run also seriously limit the effectiveness and development of therapeutic techniques. They sought to understand the particular neuroses or psychoses and, for that matter, any human being's crisis situation, not as deviations from the conceptual yardstick of this or that psychiatrist or psychologist who happened to be observing, but as deviations in the structure of that particular patient's existence, the disruption of his *condition humaine.* "A psychotherapy on existential-analytic bases investigates the life-history of the patient to be treated, . . . but it does not explain this life-history and its pathologic idiosyncrasies according to the teachings of any school of psychotherapy, or by means of its preferred categories. Instead, it *understands* this life-history as modifications of the total structure of the patient's being-in-the-world."[3]

Binswanger's own endeavor to understand how existential analysis throws light on a given case, and how it compares with other methods of psychological understanding, is graphically shown in his study of "Ellen West."[4] After he had completed his book on existential analysis, in 1942, Binswanger went back into the archives in the sanatorium of which he was director to select the case history of this young woman who had ultimately committed suicide. This case comes from 1918, before shock therapy, when psychoanalysis was in its relatively youthful stage and when the understanding of mental illness seems crude to us today. Binswanger uses the case in his endeavor to contrast the crude methods of that day with the way Ellen West would have been understood by existential psychotherapy.

Ellen West had been a tomboy in her youth and had early developed a great ambition as shown in the phrase which she used, "Either Caesar or nothing." In her late teens there becomes evident her perpetual and all-encompassing dilemmas which trapped her like vices; she vassilated from despair to joy, from anger to docility, but most of all from gorging food to starving herself. Binswanger points out the one-sidedness of the understanding of the two psychoanalysts whom Ellen West had seen, one for five months and the other for a lesser time. They interpreted her only in the world of instincts, drives, and other aspects of what Binswanger calls the *Umwelt* (to be discussed in Chapter 9). He especially takes issue with the principle, stated by Freud, in a literal translation, "In our view, perceived (observed) phenomena must yield their place to merely postulated (assumed) strivings (tendencies)."[5]

In Ellen's long illness, which we would term in our day severe anorexia nervosa, she was also seen for consultation by two psychiatrists of that day, Kraepelin, who diagnosed her as in "melancholia," and Bleuler, who offered the diagnosis of "schizophrenia."

Binswanger is not interested here in the technique of treatment but he is concerned with trying to understand Ellen West. She fascinates him by seeming to be "in love with death." In her teens Ellen implores the "Sea-King to kiss her to death." She writes, "Death is the greatest happiness in life, if not the only one" (p. 143). "If he [death] makes me wait much longer, the great friend, death, then I shall set out to seek him" (p. 242). She writes time and again that she would like to die "as the bird dies which bursts its throat in supreme joy."

Her talent as a writer is shown in her extensive poetry, diaries, and prose about her illness. She reminds one of Sylvia Plath. Binswanger poses the difficult question: Are there some persons who can fulfill their existence only by taking their own lives? "But where the existence can exist only by relinquishing life, there the existence is a tragic existence."

Ellen West seems to Binswanger to be a vivid example of Kierkegaard's description of despair in "Sickness unto Death." Binswanger writes:

> To live in the face of death, however, means "to die unto death," as Kierkegaard says; or to die one's own death, as Rilke and Scheler express it. That every passing away, every dying, whether self-chosen death or not, is still an "autonomous act" of life has already been expressed by Goethe. As he said of Raphael or Kepler, "both of them suddenly put an end to their lives," but in saying so he meant their involuntary death, coming to them "from the outside" "as external fate," so we may conversely designate Ellen West's self-caused death as a passing away or dying. Who will say where in this case guilt begins and "fate" ends?[6]

Whether or not Binswanger is successful in explicating existential principles in this case is for the reader to judge. But anyone who reads this long case will feel the amazing depth of Binswanger's earnestness in his search together with his rich cultural background and scholarliness.

It is relevant here to note the long friendship between Binswanger and Freud, a relationship which both greatly valued.

In a small book giving his recollections of Freud, which he published at the urging of Anna Freud, Binswanger recounts the many visits he made to Freud's home in Vienna and the visit of several days Freud made to him at his sanatorium on Lake Constance. Their relationship was the more remarkable since it was the sole instance of a lasting friendship of Freud with any colleague who differed radically with him. There is a poignant quality in a message Freud wrote to Binswanger in reply to the latter's New Year's letter: "You, quite different from so many others, have not let it happen that your intellectual development—which has taken you further and further away from my influence—should destroy our personal relations, and you do not know how much good such fineness does to one."[7] Whether the friendship survived because the intellectual conflict between the two was like the proverbial battle between the elephant and the walrus, who never met on the same ground, or because of some diplomatic attitude on Binswanger's part (a tendency for which Freud mildly chided him at one point) or because of the depth of their respect and affection for each other, we cannot judge. What was certainly important, however, was the fact that Binswanger and the others in the existential movement in therapy were concerned not with arguing about specific dynamisms as such but with analyzing the underlying assumptions about human nature and arriving at a *structure* on which all specific therapeutic systems could be based.

It would be a mistake, therefore, simply to identify the existential movement in psychotherapy as another in the line of schools which have broken off from Freudianism, from Jung and Adler on down. Those previous deviating schools, although called forth by blind spots in orthodox therapy and typically emerging when orthodoxy had struck an arid plateau, were nevertheless formed under the impetus of the creative work of one seminal leader. Otto Rank's new emphasis on the *present time* in the patient's experience emerged in the early 1920s, when classical analysis was bogging down in arid intellectualized discussion of the patient's past; Wilhelm Reich's *character analysis* arose in the late 1920s as an answer to the special need to break through the "ego defenses" of the character armor; new *cultural approaches* developed in the 1930s through the work of Horney and, in their distinctive ways, Fromm and Sullivan, when orthodox analysis was missing the real significance of the social and interpersonal aspects of neurotic and psychotic disturbances. Now the emergence of the existential therapy movement does have one feature in common with these other schools—namely, that it was also called forth by blind spots, as we shall make clearer later, in the existing approaches to psychotherapy. But it differs from the other schools in two respects. First, it is not the creation of any one leader, but grew up spontaneously and indigenously in diverse parts of Europe. Second, it does not purport to found a new school as opposed to other schools or to give a new technique of therapy as opposed to other techniques. It seeks, rather, to analyze the structure of human existence—an enterprise which, if successful, should yield an understanding of the reality underlying all situations of human beings in crises.

Thus this movement purports to do more than cast light upon blind spots. When Binswanger writes "existential analysis is able to widen and deepen the basic concepts and understandings of psychoanalysis," he is on sound ground, in my judgment, not only with respect to analysis but other forms of therapy as well.

When existential psychotherapy was first introduced in the United States by the book *Existence,* there was a good deal of resistance to the movement, despite the fact that it had been prominent in Europe for some time. While most of this opposition has subsided, it is still valuable to look at the nature of those resistances.

The *first* source of resistance to this or any new contribution is the assumption that all major discoveries have been made in these fields and we need only fill in the details. This attitude is an old interloper, an uninvited guest who has been notoriously present in the battles between the schools in psychotherapy. Its name is "blind-spots-structuralized-into-dogma." And though it does not merit an answer, nor is it susceptible to any, it is unfortunately an attitude which may be more widespread in this historical period than one would like to think.

The *second* source of resistance, and one to be answered seriously, is the suspicion that existential analysis is an encroachment of philosophy into psychiatry, and does not have much to do with science. This attitude is partly a hangover of the culturally inherited scars from the battle of the last of the nineteenth century, when psychological science won its freedom from metaphysics. The victory then achieved was exceedingly important, but, as in the aftermath of any war, there followed reactions to opposite extremes which are themselves harmful. Concerning this resistance we shall make several comments.

It is well to remember that the existential movement in psychiatry and psychology arose precisely out of a passion to be not *less* but *more* empirical. Binswanger and the others were convinced that the traditional scientific methods not only did not do justice to the data but actually tended to hide rather than reveal what was going on in the patient. The existential analysis movement is a protest against the tendency to see the patient in forms tailored to our own preconceptions or to make him over into the image of our own predilections. In this respect existential psychology stands squarely within the scientific tradition in its widest sense. But it broadens its knowledge of man by historical perspective and scholarly depth, by accepting the facts that human beings reveal themselves in art and literature and philosophy, and by profiting from the insights of the particular cultural movements which express the anxiety and conflicts of contemporary man.

It is also important here to remind ourselves that every

scientific method rests upon philosophical presuppositions. These presuppositions determine not only how much reality the observer with this particular method can see—they are indeed the spectacles through which he perceives—but also whether or not what is observed is pertinent to real problems and, therefore, whether the scientific work will endure. It is a gross, albeit common, error to assume naïvely that one can observe facts best if one avoids all preoccupation with philosophical assumptions. All he does, then, is mirror uncritically the particular parochial doctrines of his own limited culture. The result in our day is that science gets identified with methods of *isolating* factors and observing them from an allegedly *detached base*—a particular method which arose out of the split between subject and object made in the seventeenth century in Western culture and then developed into its special compartmentalized form in the late nineteenth and twentieth centuries. We in our day are no less subject to "methodolatry" than are members of any other culture. But it seems especially a misfortune that our understanding in such a crucial area as the psychological study of man, with the understanding of emotional and mental health depending upon it, should be curtailed by uncritical acceptance of limited assumptions. Helen Sargent has sagely and pithily remarked, "Science offers more leeway than graduate students are permitted to realize."[8]

Is not the essence of science the assumption that *reality is lawful* and, therefore, understandable, and is it not an inseparable aspect of scientific integrity that any method continuously criticize its own presuppositions? The only way to widen one's "blinders" is to analyze one's philosophical assumptions. In my judgment it is very much to the credit of the psychiatrists and psychologists in this existential movement that they seek to clarify their own bases. This enables them, as Henri Ellenberger points out,[9] to see their human subjects with a fresh clarity and to shed original light on many facets of psychological experience.

The *third* source of resistance, and to my mind the most crucial of all, is the tendency in this country to be preoccupied with technique and to be impatient with endeavors to search below such considerations to find the foundations upon which all techniques must be based. This tendency can be well explained in terms of our American social background, particularly our frontier history, and it can be well justified as our optimistic, activistic concern for helping and changing people. Our genius in the field of psychology has been until recently in the behavioristic, clinical, and applied areas, and our special contributions in psychiatry have been in drug therapy and other technical applications. Gordon Allport has described the fact that American and British psychology (as well as general intellectual climate) has been Lockean—that is, pragmatic—a tradition fitting behaviorism, stimulus and response systems, and animal psychology. The Lockean tradition, Allport points out, consists of an emphasis on the mind as *tabula rasa* on which experience writes all that is later to exist therein, whereas the Leibnitzian tradition views the mind as having a potentially active core of its own. The continental tradition, in contrast, has been Leibnitzian.[10] Now it is sobering to remind oneself that every new theoretical contribution in the field of psychotherapy until a decade ago, which has had the originality and germinating power to lead to the developing of a new school has come from continental Europe with only two exceptions—and, of these, one was grandsired by a European-born psychiatrist.[11] In this country we tend to be a nation of practitioners; but the disturbing question is: Where shall we get *what* we practice? Until recently, in our preoccupation with technique, laudable enough in itself, we have tended to overlook the fact that *technique emphasized by itself in the long run defeats even technique.*

These resistances we have named, in my judgment, far from undermining the contribution of existential analysis, demonstrate precisely its potential importance to our thinking. Despite its difficulties—due partly to its language, partly to the complexity of its thought—I believe that it is a contribution of significance and originality meriting serious study.

WHAT IS EXISTENTIALISM?

We must now remove a major stumbling block—namely, the confusion surrounding the term *existentialism.* The word is bandied about to mean everything from the posturing defiant dilettantism of some members of the avant-garde on the Left Bank in Paris, to a philosophy of despair advocating suicide, to a system of antirationalist German thought written in a language so esoteric as to exasperate any empirically minded reader. Existentialism, rather, is an expression of profound dimensions of the modern emotional and spiritual temper and is shown in almost all aspects of our culture. It is found not only in psychology and philosophy but in art—*vide* Van Gogh, Cézanne, and Picasso—and in literature—*vide* Dostoevski, Baudelaire, Kafka, and Rilke. Indeed, in many ways it is the unique and specific portrayal of the psychological predicament of contemporary Western man. This cultural movement, as we shall see, has its roots in the same historical situation and the same psychological crises which called forth psychoanalysis and other forms of psychotherapy.

Confusions about the term occur even in usually highly literate places. The *New York Times,* in a report commenting on Sartre's denunciation of, and final break with, the Russian Communists for their suppression of freedom in Hungary, identified Sartre as a leader in "existentialism, a broadly materialistic form of thought." The report illustrates two reasons for the confusion—first, the identification of existentialism in the popular mind in this country with the writings of Jean-Paul Sartre. Quite apart from the fact that Sartre is known here for his dramas, movies, and novels rather than for his major, penetrating psychological analyses, it must be emphasized that he represents a subjectivist extreme in existentialism which invites misunderstanding, and his position is by no means the most useful introduction to the movement. But the second more serious confusion in the *Times* report is its definition of existentialism as "broadly materialistic." Nothing could be less accurate—nothing, unless it be the exact opposite, namely, describing it as an idealistic form of thinking. For the very essence of this approach is that it seeks to analyze and portray the human being—whether in art or literature or philosophy or psychology—on a level which undercuts the old dilemma of materialism versus idealism.

Existentialism, in short, is the endeavor to understand man by cutting below the cleavage between subject and object which has bedeviled Western thought and science since shortly after the Renaissance. This cleavage Binswanger calls "the cancer of all psychology up to now . . . the cancer of the doctrine of subject-object cleavage of the world." The existential way of understanding human beings has some illustrious progenitors in Western history, such as Socrates in his dialogues, Augustine in his depth-psychological analyses of the self, Pascal in his struggle to find a place for the "heart's reasons which the reason knows not of." But it arose specifically just over a hundred years ago in Kierkegaard's violent protest against the reigning rationalism of his day, Hegel's "totalitarianism of reason," to use Maritain's phrase. Kierkegaard proclaimed that Hegel's identification of abstract truth with reality was an illusion and amounted to trickery. "Truth exists," wrote Kierkegaard, "only as the individual himself produces it in action." He and the existentialists following him protested firmly against the rationalists and idealists who would see man only as a subject—that is, as having reality only as a thinking being. But just as strongly they fought against the tendency to treat man as an object to be calculated and controlled, exemplified in the almost overwhelming tendencies in the Western world to make human beings into anonymous units to fit like robots into the vast industrial and political collectivisms of our day.

These thinkers sought the exact opposite of intellectualism for its own sake. They would have protested more violently than classical psychoanalysts against the use of thinking as a defense against vitality or as a substitute for immediate experience. One of the early existentialists of the sociological wing, Ludwig Feuerbach, makes this appealing admonition, "Do not wish to be a philosopher in contrast to being a man . . . do not think as a thinker . . . think as a living, real being. Think in Existence."[12]

The term *existence,* coming from the root *ex-sistere,* means literally "to stand out, to emerge." This accurately indicates what these cultural representatives sought, whether in art or philosophy or psychology—namely, to portray the human being not as a collection of static substances or mechanisms or patterns but rather as emerging and becoming, that is to say, as existing. For no matter how interesting or theoretically true is the fact that I am composed of such and such chemicals or act by such and such mechanisms or patterns, the crucial question always is that I happen to exist at this given moment in time and space, and my problem is how I am to be aware of that fact and what I shall do about it. As we shall see later, the existential psychologists and psychiatrists do not rule out the study of dynamisms, drives, and patterns of behavior. But they hold that these cannot be understood in any given person except in the context of the overarching fact that here is a person who happens *to exist, to be,* and if we do not keep this in mind, all else we know about this person will lose its meaning. Thus the existentialists approach is always dynamic; existence refers to coming into being, becoming. Their endeavor is to understand this becoming not as a sentimental artifact but as the fundamental structure of human existence. When the term *being* is used in the following pages, the reader should remember that it is not a static word but a verb form, the participle of the verb *to be.* Existentialism is basically concerned with *ontology*—that is, the science of being (*ontos,* from Greek "being").

We can see more clearly the significance of the term if we recall that traditionally in Western thought "existence" has been set over against "essence." Essence refers to the greenness of this stick of wood, let us say, and its density, weight, and other characteristics which give it substance. By and large Western thought since the Renaissance has been concerned with essences. Traditional science seeks to discover such essences or substances; it assumes an *essentialist* metaphysics, as Professor Wild puts it.[13] The search for essences may indeed produce higly significant universal laws in science or brilliant abstract conceptualizations in logic or philosophy. But it can do this only by abstraction. The *existence* of the given individual thing has to be left out of the picture. For example, we can demonstrate that three apples added to three make six. But this would be just as true if we substituted unicorns for apples; it makes no difference to the mathematical truth of the proposition whether apples or unicorns actually exist or not. Reality makes a difference to the person who *has* the apples—that is the *existential* side—but it is irrelevant to the truth of the mathematical proposition. For a more serious example, that all men die is a truth; and to say that such and such a percentage die at such and such ages gives a statistical accuracy to the proposition. But neither of these statements says anything about the fact which really matters most to each of us—namely, that you and I must alone face the fact that at some unknown moment in the future we shall die. In contrast to the essentialist propositions, these latter are *existential facts.*

All this is to say a proposition can be *true* without being *real.* Perhaps just because the abstract has worked so magnificently in certain areas of science, we tend to forget that it necessarily involves a detached viewpoint and that the living individual must be omitted. There remains the chasm between truth and reality. And the crucial question which confronts us in psychology and other aspects of the science of man is precisely this chasm between what is *abstractly true* and what is *existentially real* for the given living person.

Lest it seem that we are setting up an artificial, straw-man issue, let us point out that this chasm between truth and reality is openly and frankly admitted by sophisticated thinkers in behavioristic and conditioning psychology. Kenneth W. Spence, a leader of one wing of behavior theory, wrote, "The question of whether any particular realm of behavior phenomena is more real or closer to real life and hence should be given priority in investigation does not, or at least should not, arise for the psychologist *as scientist.*" That is to say, it does not primarily matter whether what is being studied is real or not. What realms, then, should be selected for study? Spence gave priority to phenomena which lend themselves "to the degrees of control and analysis necessary for the formulation of abstract laws."[14] Nowhere has our point been put more unabashedly and clearly—what can be reduced to *abstract laws* is selected, and whether what you are studying has reality or not is irrelevant to this goal. On the basis of this approach many an impressive system in psychology has been erected, with abstraction piled high upon abstraction—the authors succumbing, as we intellectuals are wont, to their "edifice complex"—until an admirable and imposing structure is built. The only trouble is that the edifice has more often than not been separated from human reality in its very foundations. Now the thinkers in the existential tradition hold the exact opposite to Spence's view, and so do the psychiatrists and psychologists in the existential psychotherapy movement. They insist that it is necessary and possible to have a science of man which studies human beings in their reality.

Kierkegaard, Nietzsche, and those who followed them accurately foresaw this growing split between truth and reality in Western culture, and they endeavored to call Western man back from the delusion that reality can be comprehended in an abstracted, detached way. But though they protested vehemently against arid intellectualism, they were by no means simple activists. Nor were they antirational. Anti-intellectualism and other movements in our day which make thinking subordinate to acting must not at all be confused with existentialism. Either alternative—making man subject *or* object—results in losing the living, existing person. Kierkegaard and the existential thinkers appealed to a reality *underlying both subjectivity and objectivity.* We must not only study a person's experience as such, they held, but even more we must study the man to whom the experience is happening, the one who is doing the experiencing. They insist that "Reality or Being is not the object of cognitive experience, but is rather 'existence,' is Reality as immediately experienced, with the accent on the inner, personal character of man's immediate experience."[15] This comment, as well as several above, will indicate to the reader how close the existentialists are to present-day depth psychology. It is by no means accidental that the greatest of them in the nineteenth century, Kierkegaard and Nietzsche, happen also to be among the most remarkable psychologists (in the dynamic sense) of all time and that one of the contemporary leaders of this school, Karl Jaspers, was originally a psychiatrist and wrote a notable text on psychopathology. When one reads Kierkegaard's profound analyses of anxiety and despair or Nietzsche's amazingly acute insights into the dynamics of resentment and the guilt and hostility which accompany repressed emotional powers, one must pinch oneself to realize that one is reading works written in the last century and not some new contemporary psychological analysis. The existentialists are centrally concerned with rediscovering the living person amid the compartmentalization and dehumanization of modern culture, and in order to do this they engage in depth psychological analysis. Their concern is not with isolated psychological reactions in themselves but rather with the psychological being of the living man who is doing the experiencing. That is to say, they use psychological terms with an ontological meaning.

In the winter of 1841, Schelling gave his famous series of lectures at the University of Berlin before a distinguished audience including Kierkegaard, Burckhardt, Engels, Bakunin. Schelling set out to overthrow Hegel, whose vast rationalist system, including the identification of abstract truth with reality and the bringing of all of history into an "absolute whole," held immense and dominant popularity in the Europe of the middle nineteenth century. Though many of Schelling's listeners were bitterly disappointed in his answers to Hegel, the existential movement may be said to have begun there. Kierkegaard went back to Denmark and in 1844 published his *Philosophical Fragments,* and two years later he wrote the declaration of independence of existentialism, *Concluding Unscientific Postscript.* Also in 1844 there appeared the second edition of Schopenhauer's *The World as Will and Idea,* a work important in the new movement because of its central emphasis on vitality, "will," along with "idea." Two related works were written by Karl Marx in 1844–1845. The early Marx is significant in this movement in his attack upon abstract truth as "ideology," again using Hegel as his whipping boy. Marx's dynamic view of history as the arena in which men and groups bring truth into being and his meaningful fragments pointing out how the money economy of modern industrialism tends to turn people into things and works toward the dehumanization of modern man are likewise significant in the existentialist approach. Both Marx and Kierkegaard took over Hegel's dialectical method but used it for quite different purposes. More existential elements were latently present in Hegel, it may be noted, than his antagonists acknowledged.

In the following decades the movement subsided. Kierkegaard remained completely unknown, Schelling's work was contemptuously buried, and Marx and Feuerbach were interpreted as dogmatic materialists. Then a new impetus came in the 1880s with the work of Dilthey, and particularly with Friedrich Nietzsche, the "philosophy of life" movement, and the work of Bergson.

The third and contemporary phase of existentialism came after the shock to the Western world caused by World War I. Kierkegaard and the early Marx were rediscovered, and the serious challenges to the spiritual and psychological bases of Western society given by Nietzsche could no longer be covered over by Victorian self-satisfied placidity. The specific form of this third phase owes much to the phenomenology of Edmund Husserl, which gave to Heidegger, Jaspers, and the others the tool they needed to undercut the subject-object cleavage which had been such a stumbling block in science as well as philosophy. There is an obvious similarity between existentialism, in its emphasis on truth as produced in action, with the process philosophies, such as Whitehead's, and American pragmatism, particularly as in William James.[16]

Martin Heidegger is generally taken as the fountainhead of present-day existential thought. His seminal work, *Being and Time,* was of radical importance in giving Binswanger and other existential psychiatrists and psychologists the deep and broad basis they sought for understanding the human being. Heidegger's thought is rigorous, logically incisive, and "scientific" in the European sense of pursuing with unrelenting vigor and thoroughness whatever implications his inquiries led him to. His work is difficult to translate, but Macquarrie and Robinson succeeded with regard to *Being and Time.*[17] Jean-Paul Sartre's best contribution to our subject are his phenomenological descriptions of psychological processes. In addition to Jaspers, other prominent existential thinkers are Gabriel Marcel in France, Nicolas Berdyaev, originally Russian but until his death a resident of Paris, and Ortega y Gasset and Unamuno in Spain. Paul Tillich shows the existential approach in his work, and in many ways his book *The Courage to Be* is the best and most cogent presentation in English of existentialism as an approach to actual living.[18]

The novels of Kafka portray the despairing, dehumanized situation in modern culture from which and to which existentialism speaks. *The Stranger* and *The Plague,* by Albert Camus, represent excellent examples in modern literature in which existentialism is partially self-conscious. But perhaps the most vivid of all portrayals of the meaning of existentialism is to be found in modern art, partly because it is articulated symbolically rather than as self-conscious thought and partly because art always reveals with special clarity the underlying spiritual and emotional temper of the culture. The common elements in the work of such outstanding representatives of the modern movement as Van Gogh, Cézanne, and Picasso are, first, a revolt against the hypocritical academic tradition of the late nineteenth century; second, an endeavor to pierce below surfaces to grasp a new relation to the reality of nature; third, an endeavor to recover vitality and honest, direct aesthetic experience; and fourth, the desperate attempt to express the immediate underlying meaning of the modern human situation, even when this means portraying despair and emptiness. Tillich, for example, holds that Picasso's painting *Guernica* gives the most gripping and revealing portrayal of the atomistic, fragmentized condition of European society which preceded World War II and "shows what is now in the souls of many Americans as disruptiveness, existential doubt, emptiness and meaninglessness."[19]

The fact that the existential approach arose as an indigenous and spontaneous answer to crises in modern culture is shown not only in the fact that it emerged in art and literature but also in the fact that different philosophers in diverse parts of Europe often developed these ideas without conscious relation to each other. Though Heidegger's main work, *Being and Time,* was published in 1927, Ortega y Gasset already in 1924 had developed and partially published strikingly similar ideas in *The Dehumanization of Art, and Other Writings on Art and Culture.*[20]

It is true that existentialism had its birth in a time of cultural crisis, and it is always found in our day on the sharp revolutionary edge of modern art, literature, and thought. To my mind this fact speaks for the validity of its insights rather than the reverse. When a culture is caught in the profound convulsions of a transitional period, the individuals in the society understandably suffer spiritual and emotional upheaval; and finding that the accepted mores and ways of thought no longer yield security, they tend either to sink into dogmatism and conformism, giving up awareness, or are forced to strive for a heightened self-consciousness by which to become aware of their existence with new conviction and on new bases. This is one of the most important affinities of the existential movement with psychotherapy—both are concerned with individuals in crisis. And far from saying that the insights of a crisis period are "simply the product of anxiety and despair," we are more likely to find, as we do time and again in psychoanalysis, that a crisis is exactly what is required to shock people out of unaware dependence upon external dogma and to force them to unravel layers of pretense to reveal naked truth about themselves which, however unpleasant, will at least be solid.

Existentialism is an attitude which accepts man as always becoming, which means potentially in crisis. But this does not mean it will be despairing. Socrates, whose dialectical search for truth in the individual is the prototype of existentialism, was optimistic. But this approach is understandably more apt to appear in ages of transition, when one age is dying and the new one not yet born, and the individual is either homeless and lost or achieves a new self-consciousness. In the period of transition from medievalism to the Renaissance, a moment of radical upheaval in Western culture, Pascal describes powerfully the experience the existentialists later were to call *Dasein:* "When I consider the brief span of my life, swallowed up in the eternity before and behind it, the small space that I fill, or even see, engulfed in the infinite immensity of spaces which I know not, and which know not me, I am afraid, and wonder to see myself here rather than there; for there is no reason why I should be here rather than there, now rather than then."[21] Rarely has the existential problem been put more simply or beautifully. In this passage we see, first, the profound realization of the contingency of human life which existentialists call "thrownness." Second, we see Pascal facing unflinchingly the question of *being there* or more accurately "being where?" Third, we see the realization that one cannot take refuge in some superficial explanation of time and space, which Pascal, scientist that he was, could well know; and last, the deep shaking anxiety arising from this stark awareness of existence in such a universe.[22]

It remains, finally, to note the relation between existentialism and Oriental thought as shown in the writings of Laotzu and Zen Buddhism. The similarities are striking. One sees this immediately in glancing at some quotations from Laotzu's *The Way of Life:* "Existence is beyond the power of words to define: terms may be used but none of them is absolute." "Existence, by nothing bred, breeds everything, parent of the universe." "Existence is infinite, not to be defined; and though it seem but a bit of wood in your hand, to carve as you please, it is not to be lightly played with and laid down." "The way to do is to be." "Rather abide at the center of your being; for the more you leave it, the less you learn."[23]

One gets the same shock of similarity in Zen Buddhism.[24] The likenesses between these Eastern philosophies and existentialism go much deeper than the chance similarity of words. Both are concerned with ontology, the study of being. Both seek a relation to reality which cuts below the cleavage between subject and object. Both would insist that the Western absorption in conquering and gaining power over nature has resulted not only in the estrangement of man from nature but also indirectly in the estrangement of man from himself. The basic reason for these similarities is that Eastern thought never suffered the radical split between subject and object that has characterized Western thought, and this dichotomy is exactly what existentialism seeks to overcome.

The two approaches are not at all to be identified; they are on different levels. Existentialism is not a comprehensive philosophy or way of life, but an endeavor to grasp reality. The chief specific difference between the two, for our purposes, is that existentialism is immersed in and arises directly out of Western man's anxiety, estrangement, and conflicts and is indigenous to our culture. Like psychoanalysis, existentialism seeks not to bring in answers from other cultures but to utilize these very conflicts in contemporary personality as avenues to the more profound self-understanding of Western man and to find the solutions to our problems in direct relation to the historical and cultural crises which gave the problems birth. In this respect, the particular value of Eastern thought is not that it can be transferred, ready-born like Athena, to the Western mind, but rather that it serves as a corrective to our biases and highlights the erroneous assumptions that have led Western development to its present problems. The present widespread interest in Oriental thought in the Western world is, to my mind, a reflection of the same cultural crises, the same sense of estrangement, the same hunger to get beyond the vicious circle of dichotomies which called forth the existentialist movement.

FOUR

How Existentialism and Psychoanalysis Arose Out of the Same Cultural Situation

WE SHALL NOW look at the remarkable parallel between the problems of modern man to which the existentialists on one hand and psychoanalysts on the other devote themselves. From different perspectives and on different levels, both analyze anxiety, despair, alienation of man from himself and his society, and both seek a synthesis of integration and meaning in the person's life.

Freud describes the neurotic personality of the late nineteenth century as one suffering from fragmentation—that is, from repression of instinctual drives, blocking off of awareness, loss of autonomy, weakness and passivity of the ego, together with the various neurotic symptoms which result from this fragmentation. Kierkegaard—who wrote the only known book before Freud specifically devoted to the problem of anxiety—analyzes not only anxiety but particularly the depression and despair which result from the individual's self-estrangement, an estrangement he proceeds to classify in its different forms and degrees of severity.[1] Nietzsche proclaims, ten years before Freud's first book, that the disease of contemporary man is that "his soul had gone stale," he is "fed up," and that all about there is "a bad smell . . . the smell of failure. . . . The leveling and diminution of European man is our greatest danger." He then proceeds to describe, in terms which remarkably predict the later psychoanalytic concepts, how blocked instinctual powers turn within the individual into resentment, self-hatred, hostility, and aggression. Freud did not know Kierkegaard's work, but he regarded Nietzsche as one of the authentically great men of all time.

What is the relation between these three giants of the nineteenth century, none of whom directly influenced either of the others? And what is the relation between the two approaches to human nature they originated—existentialism and psychoanalysis—probably the two most important to have shaken, and indeed toppled, the traditional concepts of man? To answer these questions we must inquire into the cultural situation of the middle and late nineteenth century out of which both approaches to man arose and to which both sought to give answers. The real meaning of a way of understanding human beings, such as existentialism or psychoanalysis, can never be seen *in abstracto,* detached from its world, but only in the context of the historical situation which gave it birth. Thus the historical discussions to follow are not at all detours from our central aim. Indeed, it is precisely this historical approach which may throw light on our chief question—namely, how the specific scientific techniques that Freud developed for the investigation of the fragmentation of the individual in the Victorian period are related to the understanding of man and his crises to which Kierkegaard and Nietzsche contributed so much and which later provided a broad and deep base for existential psychotherapy.

COMPARTMENTALIZATION AND INNER BREAKDOWN IN THE NINETEENTH CENTURY

The chief characteristic of the last half of the nineteenth century was the breaking up of personality into fragments. These fragmentations, as we shall see, were symptoms of the emotional, psychological, and spiritual disintegration occurring in the culture and in the individual. One can see this splitting up of the individual personality not only in the psychology and the science of the period but in almost every aspect of late nineteenth-century culture. One can observe the fragmentation in family life, vividly portrayed and attacked in Ibsen's *A Doll's House.* The respectable citizen who keeps his wife and family in one compartment and his business and other worlds in others is making his home a doll's house and preparing its collapse. One can likewise see the compartmentalization in the separation of art from the realities of life, the use of art in its prettified, romantic, academic forms as a hypocritical escape from existence and nature, the art as *art*ificiality against which Cézanne, Van Gogh, the impressionists, and other modern art movements so vigorously protested. One can furthermore see the fragmentation in the separating of religion from weekday existence, making it an affair of Sundays and special observances, and the divorce of ethics from business. The segmentation was occurring also in philosophy and psychology—when Kierkegaard fought so passionately against the enthronement of an arid, abstract reason and pleaded for a return to reality, he was by no means tilting at windmills. The Victorian man saw himself as segmented into reason, will, and emotions and found the picture good. His reason was supposed to tell him *what* to do, then voluntaristic will was supposed to give him the means to do it, and emotions—well, emotions could best be channeled into compulsive business drive and rigidly structuralized in Victorian mores; and the emotions which would really have upset the formal segmentation, such as sex and hostility, were to be stanchly repressed or let out only in orgies of patriotism or on well-contained weekend "binges" in Bohemia in order that one might, like a steam engine which has let off surplus pressure, work more effectively on returning to his desk Monday morning. Naturally, this kind of man had to put great stress on "rationality." Indeed, the very term *irrational* means a thing not to be spoken of or thought of; and Victorian man's repressing, or compartmentalizing, what was not to be thought of was a precondition for the apparent stability of the culture. Schachtel has pointed out how the citizen of the Victorian period so needed to persuade himself of his own rationality that he denied the fact that he had ever been a child or had a child's irrationality and lack of control; hence the radical split between the adult and the child, which was portentous for Freud's investigations.[2]

This compartmentalization went hand in hand with the developing industrialism, as both cause and effect. A man who can keep the different segments of his life entirely separated, who can punch the clock every day at exactly the same moment, whose actions are always predictable, who is never troubled by irrational urges or poetic visions, who indeed can manipulate himself the same way he would the machine whose levers he pulls, is the most profitable worker not only on the assembly line but even on many of the higher levels of production. As Marx and Nietzsche pointed out, the corollary is likewise true: the very success of the industrial system, with its accumulation of money as a validation of personal worth entirely separate from the actual product of a man's hands, had a reciprocal depersonalizing and dehumanizing effect upon man in his relation to others and himself. It was against these dehumanizing tendencies to make man into a machine, to make him over in the image of the industrial system for which he labored, that the early existentialists fought so strongly. And they were aware that the most serious threat of all was that reason would join mechanics in sapping the individual's vitality and decisiveness. *Reason,* they predicted, *was becoming reduced to a new kind of technique.*

Scientists in our day are often not aware that this compartmentalization, finally, was also characteristic of the sciences of the century of which we are heirs. This nineteenth century was the era of the "autonomous sciences," as Ernest Cassirer phrases it. Each science developed in its own direction; there was no unifying principle, particularly with relation to man. The views of man in the period were supported by empirical evidence amassed by the advancing sciences, but "each theory became a Procrustean bed on which the empirical facts were stretched to fit a preconceived pattern. . . . Owing to this development our modern theory of man lost its intellectual center. We acquired instead a complete anarchy of thought. . . . Theologians, scientists, politicians, sociologists, biologists, psychologists, ethnologists, economists all approached the problem from their own viewpoints . . . every author seems in the last count to be led by his own conception and evaluation of human life."[3] It is no wonder that Max Scheler declared, "In no other period of human knowledge has man ever become more problematic to himself than in our own days. We have a scientific, a philosophical, and a theological anthropology that know nothing of each other. Therefore we no longer possess any clear and consistent idea of man. The ever-growing multiplicity of the particular sciences that are engaged in the study of men has much more confused and obscured than elucidated our concept of man."[4]

On the surface the Victorian period appeared placid, contented, ordered; but this placidity was purchased at the price of widespread, profound, and increasingly brittle repression. As in the case of an individual neurotic, the compartmentalization became more and more rigid as it approached the point—August 11, 1914—when it was to collapse altogether.

Now it is to be noted that the compartmentalization of the

culture had its *psychological parallel in radical repression within the individual personality.* Freud's genius was in developing scientific techniques for understanding, and mayhap curing, this fragmentized individual personality; but he did not see —until much later, when he reacted to the fact with pessimism and some detached despair[5]—that the neurotic illness in the individual was only one side of disintegrating forces which affected the whole of society. Kierkegaard, for his part, foresaw the results of this disintegration upon the inner emotional and spiritual life of the individual: endemic anxiety, loneliness, estrangement of one man from another, and finally the condition that would lead to ultimate despair, man's alienation from himself. But it remained for Nietzsche to paint most graphically the approaching situation: "We live in a period of atoms, of atomic chaos," and out of this chaos he foresaw, in a vivid prediction of collectivism in the twentieth century, "the terrible apparition . . . the Nation State . . . and the hunt for happiness will never be greater than when it must be caught between today and tomorrow; because the day after tomorrow all hunting time may have come to an end altogether. . . ."[6] Freud saw this fragmentation of personality in the light of natural science and was concerned with formulating its technical aspects. Kierkegaard and Nietzsche did not underestimate the importance of the specific psychological analysis; but they were much more concerned with understanding *man as the being who represses,* the being who surrenders self-awareness as a protection against reality and then suffers the neurotic consequences. The strange question is: What does it mean that man, the being in the world who can be conscious that he exists and can know his existence, should choose or be forced to choose to block off this consciousness and should suffer anxiety, compulsions for self-destruction, and despair? Kierkegaard and Nietzsche were keenly aware that the "sickness of soul" of Western man was a deeper and more extensive morbidity than could be explained by the specific individual or social problems.

Something was radically wrong in man's relation to himself; man had become fundamentally problematic to himself. "This is Europe's true predicament," declared Nietzsche; "together with the fear of man we have lost the love of man, confidence in man, indeed, *the will to man.*"

FIVE

Kierkegaard, Nietzsche, and Freud

WE TURN NOW to a more detailed comparison of the approach to understanding Western man given by Kierkegaard and Nietzsche, with the hope of seeing more clearly their interrelationship with the insights and methods of Freud.

KIERKEGAARD

Kierkegaard's penetrating analysis of anxiety—which we have summarized in another volume[1]—would alone assure him of a position among the psychological geniuses of all time. His insights into the significance of self-consciousness, his analysis of inner conflicts, loss of the self, and even psychosomatic problems are the more surprising since they antedate Nietzsche by four decades and Freud by half a century. This indicates in Kierkegaard a remarkable sensitivity to what was going on under the surface of Western man's consciousness in his day, to erupt only half a century later. He died at the early age of forty-four, after an intense, passionate, and lonely period of creativity in which he wrote almost two dozen books in the space of fifteen years. Secure in the knowledge that he would become important in decades to come, he had no illusions about his discoveries and insights being welcomed in his day. "The present writer," he says in one satirical passage about himself, "is nothing of a philosopher; he is . . . an amateur writer who neither writes the System nor promises the System nor ascribes anything to it. . . . He can easily foresee his fate in an age when passion has been obliterated in favor of learning, in an age when an author who wants to have readers must take care to write in such a way that the book can easily be perused during the afternoon nap. . . . He foresees his fate, that he will be entirely ignored." True to his prediction, he was almost unknown in his day—except for satirical lampooning in *Corsair,* the humor magazine of Copenhagen. For half a century he remained forgotten and was then rediscovered in the second decade of this century, not only to have a profound effect on philosophy and religion but also to yield specific and important contributions to depth-psychology. Binswanger, for example, states in his paper on Ellen West that she "suffered from that sickness of the mind which Kierkegaard, with the keen insight of genius, described and illuminated from all possible aspects under the name of 'Sickness Unto Death.' I know of no document which could more greatly advance the existential-analytic interpretation of schizophrenia than that. One might say that in this document Kierkegaard had recognized with intuitive genius the coming of schizophrenia. . . ." Binswanger goes on to remark that the psychiatrist or psychologist who does not concur in Kierkegaard's religious interpretations nevertheless remains "deeply indebted to this work of Kierkegaard."

Kierkegaard, like Nietzsche, did not set out to write philosophy or psychology. He sought only to understand, to uncover, to disclose human existence. With Freud and Nietzsche he shared a significant fact: all three of them based their knowledge chiefly on the analysis of one case—namely, themselves. Freud's germinal books, such as *Interpretation of Dreams,* were based almost entirely on his own experience and his own dreams; he wrote in so many words to Fliess that the case he struggled with and analyzed continually was himself. Every system of thought, remarked Nietzsche, "says only: this is a picture of all life, and from it learn the meaning of your life. And conversely; read only your life and understand from it the hieroglyphics of universal life."[2]

The central psychological endeavor of Kierkegaard may be summed up under the heading of the question he pursued relentlessly: How can you become an individual? The individual was being swallowed up on the rational side by Hegel's vast logical "absolute Whole," on the economic side by the increasing objectification of the person, and on the moral and spiritual side by the soft and vapid religion of his day. Europe was ill, and was to become more so, not because knowledge or techniques were lacking but because of the want of *passion, commitment.*[3] "Away from Speculation, away from the System," he called, "and back to reality!" He was convinced not only that the goal of "pure objectivity" is impossible but that even if it were possible it would be undesirable. And from another angle it is immoral: we are so involved in each other and the world that we cannot be content to view truth disinterestedly. Like all the existentialists, he took the term *interest* (inter-est) seriously.[4] Every question is the "question for the Single One" —that is, for the alive and self-aware individual; and if we don't start with the human being there, we shall have spawned, with all our technical prowess, a collectivism of robots who will end up not just in emptiness but in self-destructive despair.

One of the most radical contributions of Kierkegaard to later dynamic psychology is his formulation of truth as relationship. In the book which was later to become the manifesto for existentialism, he writes:

When the question of truth is raised in an objective manner, reflection is directed objectively to the truth, as an object to which the knower is related. Reflection is not focused upon the relationship, however, but upon the question of whether it is the truth to which the knower is related. If only the object to which he is related is the truth, the subject is accounted to be in the truth. *When the question of the truth is raised subjectively, reflection is directed subjectively to the nature of the individual's relationship; if only the mode of this relationship is in the truth, the individual is in the truth, even if he should happen to be thus related to what is not true.* [5]

It would be hard to exaggerate how revolutionary these sentences were and still are for modern culture as a whole and for psychology in particular. Here is the radical, original statement of *relational truth.* Here is the fountainhead of the emphasis in existential thought on truth as *inwardness* or, as Heidegger puts it, truth as freedom.[6] Here, too, is the prediction of what was later to appear in twentieth-century physics—namely, the reversal of the principle of Copernicus that one discovered truth most fully by detaching man, the observer. Kierkegaard foretells the viewpoint of Bohr, Heisenberg, and other modern physicists that the Copernican view that nature can be separated from man is no longer tenable. The "ideal of a science which is completely independent of man [i.e., completely objective] is an illusion," in Heisenberg's words.[7] Here is, in Kierkegaard's paragraph, the forerunner of relativity and the other viewpoints which affirm that the human being who is engaged in studying the natural phenomena is in a particular and significant relationship to the objects studied and he must make himself part of his equation. That is to say, the *subject,* man, can never be separated from the *object* which he observes. It is clear that the cancer of Western thought, the subject-object split, received a decisive attack in this analysis of Kierkegaard's.

But the implications of this landmark are even more specific and more incisive in psychology. It releases us from bondage to the dogma that truth can be understood only in terms of external *objects.* It opens up the vast provinces of inner, subjective reality and indicates that such reality may be true even though it contradicts objective fact. This was the discovery Freud was later to make when, somewhat to his chagrin, he learned that the "childhood rape" memories so many of his patients confessed were generally lies from a factual point of view, the rape never having in fact occurred. But it turned out that the experience of rape was as potent even if it *existed only in fantasy,* and that in any case the crucial question was how the patient *reacted to* the rape rather than whether it was true or false in fact. We have, thus the opening of a continent of new knowledge about inner dynamics when we take the approach that the *relation to* a fact or person or situation is what is significant for the patient or person we are studying and the question of whether or not something objectively occurred is on a quite different level. Let us, to avoid misunderstanding, emphasize even at the price of repetition that this truth-as-relationship principle does not in the slightest imply a sloughing off of the importance of whether or not something is objectively true. This is not the point. Kierkegaard is not to be confused with the subjectivists or idealists; he opens up the subjective world without losing objectivity. Certainly one has to deal with the real objective world; Kierkegaard, Nietzsche, and their ilk took nature more seriously than many who call themselves naturalists. The point rather is that the meaning for the person of the objective fact (or fantasied one) depends on how he relates to it; there is no existential truth which can omit the relationship. An objective discussion of sex, for example, may be interesting and instructive; but once one is concerned with a given person, the objective truth depends for its meaning upon the relationship between that person and the sexual partner, and to omit this factor not only constitutes an evasion but cuts us off from seeing reality.

The approach stated in Kierkegaard's sentences is, furthermore, the forerunner of concepts of "participant observation" of Sullivan and the other emphases upon the significance of the therapist in the relationship with the patient. The fact that the therapist participates in a real way in the relationship and is an inseparable part of the "field" does not, thus, impair the soundness of his scientific observations. Indeed, can we not assert that unless the therapist is a real participant in the relationship and consciously recognizes this fact, he will *not* be able to discern with clarity what is in fact going on? The implication of this "manifesto" of Kierkegaard is that we are freed from the traditional doctrine, so limiting, self-contradictory, and indeed often destructive in psychology, *that the less we are involved in a given situation, the more clearly we can observe the truth.* The implication of that doctrine was, obviously enough, that there is an inverse relation between involvement and our capacity to observe without bias. And the doctrine became so well enshrined that we overlooked another one of its clear implications —namely, that he will most successfully discover truth who is not the slightest bit interested in it! No one would argue against the obvious fact that *disruptive* emotions interfere with one's perception. In this sense it is self-evident that anyone in a therapeutic relationship, or any person observing others, for that matter, must clarify very well what his particular emotions and involvement are in the situation. But the problem cannot be solved by detachment and abstraction. That way we end up with a handful of sea foam, and the reality of the person has evaporated before our eyes. The clarification of the pole in the relationship represented by the therapist can only be accomplished by a fuller awareness of the existential situation—that is, the real, living relationship.[8] When we are dealing with human beings, no truth has reality by itself; it is always dependent upon the reality of the immediate relationship.

A second important contribution of Kierkegaard to dynamic psychology lies in his emphasis upon the necessity of commitment. This follows from the points already made above. Truth becomes reality only as the individual produces it in action, which includes producing it in his own consciousness. Kierkegaard's point has the radical implication that we cannot even *see* a particuar truth unless we already have some commitment to it. It is well known to every therapist that patients can talk theoretically and academically from now till doomsday about their problems and not really be affected; indeed, particularly in cases of intellectual and professional patients, this very talking, though it may masquerade under the cloak of unbiased and unprejudiced inquiry into what is going on, is often the defense against seeing the truth and against committing oneself, a defense indeed against one's own vitality. The patient's talking will not help him to get to the reality until he can experience something or some issue in which he has an immediate and absolute stake. This is often expressed under the rubric of "the necessity of arousing anxiety in the patient." I believe, however, that this puts the matter too simply and partially. Is not the more fundamental principle that the patient must find or discover some point in his existence where he can commit himself before he can permit himself even to see the truth of what he is doing? This is what Kierkegaard means by "passion" and "commitment" as over against objective disinterested observation. One corollary of this need for commitment is the commonly accepted phenomenon that we cannot get to the underlying levels of a person's problems by laboratory experimentation; only when the person himself has some hope of getting relief from his suffering and despair and of receiving some help in his problems will he undertake the painful process of investigating his illusions and uncovering his defenses and rationalizations.

NIETZSCHE

We turn now to Friedrich Nietzsche (1844–1900). He was very different from Kierkegaard in temperament, and, living four decades later, he reflected nineteenth-century culture at a different stage. He never read Kierkegaard; his friend Brandes called his attention to the Dane two years before Nietzsche's death, too late for Nietzsche to know the works of his predecessor, who was superficially so different but in many essentials so alike. Both represent in fundamental ways the emergence of the existential approach to human life. Both are often cited together as the thinkers who discerned most profoundly and predicted most accurately the psychological and spiritual state of Western man in the twentieth century. Like Kierkegaard, Nietzsche was not antirational, nor should he be confused with the "philosophers of feeling" or the "back to nature" evangelists. He attacked not reason but *mere* reason, and he attacked it in the arid, fragmentized, rationalistic form it assumed in his day. He sought to push reflection—again like Kierkegaard—to its uttermost limits to find the reality which underlies *both* reason and unreason. For reflection is, after all, a turning in on itself, a mirroring, and the issue for the living existential person is *what* he is reflecting; otherwise reflecting empties the person of vitality.[9] Like the depth psychologists to follow him, Nietzsche sought to bring into the scope of existence the unconscious, irrational sources of man's power and greatness as well as his morbidity and self-destructiveness.

Another significant relationship between these two figures and depth psychology is that they both developed a great intensity of self-consciousness. They were well aware that the most devastating loss in their objectivating culture was the individual's consciousness of himself—a loss to be expressed later in Freud's symbol of the ego as weak and passive, "lived by the Id," having lost its own self-directive powers.[10] Kierkegaard had written, "the more consciousness, the more self," a statement which Sullivan was to make in a different context a century later and which is implied in Freud's description of the aim of his technique as the increasing of the sphere of consciousness: "Where Id was, there ego shall be." But Kierkegaard and Nietzsche could not escape, in their special historical situations, the tragic consequences of their own intensity of self-consciousness. Both were lonely, anticonformist in the extreme, and knew the deepest agonies of anxiety, despair, isolation. Hence they could speak from an immediate personal knowledge of these ultimate psychological crises.[11]

Nietzsche formulated the truths that we all struggle toward. He lived and wrote at a time—the last half of the nineteenth century—when European man was psychologically and spiritually disintegrating. Outwardly the period was still one of stability and bourgeois conformism. But inwardly the spiritual rotting of human beings (if I may take my cue from Nietzsche's own words) was visible to Nietzsche. Religious faith had been transformed into resentment, vitality into sexual repression, and a general hypocrisy marked the condition of man at that time.

In an age like that one, and like our own, one had to be a psychologist in order to be a good philosopher. For man was crying to be ministered to—man who had lost his center, who was suffering psychological and spiritual disorientation.

Nietzsche was brilliantly fitted for being the physician to this disoriented man. He often spoke of himself as a "psychologist." In *Beyond Good and Evil* he argued "that psychology should be recognized again as the queen of the sciences, for whose service and preparation the other sciences exist. For psychology is now again the path to the fundamental problems."

Nietzsche's concept of "superman" and "will to power" are endeavors to rediscover some fiber, some basis for strength in his contemporaries. The widespread view that Nietzsche was nihilistic, that he was the enemy of religion, the enemy of morality and almost everything else, is a radical misunderstanding. It occurs because of a failure to see not only the meaning of Nietzsche himself, but of the world which called Nietzsche forth and to which he spoke.

Nietzsche held that one should experiment on all truth not simply in the laboratory but in one's own experience; every truth should be faced with the question, "Can one live it?" "All truths," he put it, "are bloody truths for me." Hence his famous phrase, "error is cowardice." In taking religious leaders to task for their being alien to intellectual integrity, he charges that they never make "their experiences a matter of conscience for knowledge. 'What have I really experienced? What happened then in me and around me? Was my reason bright enough? Was my will turned against all deceptions . . . ?' thus none of them questioned. . . . We, however, we others who thirst for reason want to look our experiences in the eye as severely as at a scientific experiment . . . ! We ourselves want to be our experiments and guinea-pigs!"[12] Neither Kierkegaard nor Nietzsche had the slightest interest in starting a movement —or a new system, a thought which would indeed have offended them. Both proclaimed, in Nietzsche's phrase, "Follow not me, but you!"

Both were aware that the psychological and emotional disintegration which they described as endemic, if still underground, in their periods was related to man's loss of faith in his essential dignity and humanity. Here they expressed a "diagnosis" to which very little attention was paid among the schools of psychotherapy until the past decade, when man's loss of faith in his own dignity began to be seen as a real and serious aspect of modern problems. This loss, in turn, was related to the breakdown of the convincing and compelling power of the two central traditions which had given a basis for values in Western society—namely the Hebrew-Christian and the humanistic. Such is the presupposition of Nietzsche's powerful parable "God Is Dead." Kierkegaard had passionately denounced, with almost nobody listening, the softened, vapid, and anemic trends in Christianity; by Nietzsche's time the deteriorated forms of theism and emotionally dishonest religious practices had become part of the illness and had to die. Roughly speaking, Kierkegaard speaks out of a time when God is dying, Nietzsche when God is dead. Both were radically devoted to the nobility of man and both sought some basis on which this dignity and humanity could be re-established. This is the meaning of Nietzsche's "man of power" and Kierkegaard's "true individual."

One of the reasons Nietzsche's influence upon psychology and psychiatry has so far been unsystematic, limited to a chance quotation of an aphorism here and there, is precisely that his mind is so unbelievably fertile, leaping incredibly from insight to flashing insight. The reader must take care not to be carried away in uncritical admiration or, on the other hand, to overlook Nietzsche's real importance because the richness of his thought beggars all our tidy categories. Hence we shall here endeavor briefly to clarify more systematically some of his central points.

The first essential thing about Nietzsche's philosophy is that he uses psychological terms with an ontological meaning. He shares this characteristic with his fellow existentialists, such as Kierkegaard, Sartre, Heidegger. Despair, will, anxiety, guilt, loneliness—these normally refer to psychological conditions, but for Nietzsche they refer to states of being. Anxiety, for example, is not an "affect" that you can feel at some times and not at other times. It refers rather to a state of existence. It is not something we "have," but something we "are."

The same holds for will. The term *will* in Nietzsche also refers to a basic feature of our existence. It is potentially present at all times; without it we would not be human beings. The acorn becomes an oak regardless of any choice, but man cannot realize his being except as he wills it in his encounters. In animals and plants, nature and being are one, but in man, nature and being are never to be identified. Nietzsche heaps scorn on those who still suffer under this illusion and who want to live simply according to nature. In *Beyond Good and Evil,* he cries, "According to nature you want to live? Oh, you noble stoics. What deceptive words these are. Imagine a being like nature, wasteful beyond measure, indifferent beyond measure, without purpose and consideration, without mercy and justice, fertile and desolate and uncertain at the same time. Imagine indifference itself as a power. How could you live according to this indifference?" Human values are not given us by nature but are set for us as tasks to be achieved. It is clear that using these psychological terms with an ontological significance gives them a much more profound and powerful meaning.

I believe this use of psychological terms with ontological meaning occurs in all periods when values are in transition. It certainly occured in Greek history in the first and second centuries B.C. If I alone am in despair, I may be upset by it but I can look around at others who are not in despair. That's some comfort. But if everyone is in despair, if society is in radical transition, then we are in despair en masse. Then it is a different thing; we have no north star to steer by. If our anxiety is not blocked by apathy, it tends to move into panic. This very apathy is a defense against the panic which would occur if one really did let oneself feel. We are then in a condition somewhat like Bosch's paintings of hell. Every mooring place is gone, and human beings in droves are herded into the fires. In such times the ontological use of psychological terms represents an endeavor to gain some new basis, some new foundation for our values. This is what Nietzsche was struggling for.

Finally, there is the question of the contribution to psychology of Nietzsche's concept of power, and specifically the "will to power." "The near-most essence of Being," Nietzsche writes, "is will to power." In academic psychology not only is this idea of Nietzsche's not recognized, but the concept of power itself is completely repressed. Sometimes, to be sure, it is subsumed under "will," but "will" also has been largely ignored since the days of William James.

I think the general repression of power and of the topic of power has been lamentable in the extreme. The choice of substitute terms is very revealing. Take for example, the concept of "control." Control is a substitute expression for power, which puts the accent on *my* right to exert power over *you*. It would be clearer to use the word *power* to begin with.

What is the meaning of "power" in the development of this concept? The phrase "will to power" means self-actualization. Nietzsche was protesting against the weak, anemic, European man he saw emerging on all sides. The "will to power" is a call to man to avoid the putrescense and to affirm himself in his existence with strength and commitment. The "will to power" is built into every individual because it is inseparable from life itself. "Wherever I found life," writes Nietzsche, "there I found the will to power."

His concept of "will to power" implies the self-realization of the individual in the fullest sense. It requires the courageous living out of the individual's potentialities in his own particular existence. Like all existentialists, Nietzsche is not using psychological terms to describe psychological attributes or faculties or a simple pattern of behavior, such as aggression or power over someone. Will to power rather is an ontological category—that is to say, an inseparable aspect of being. It does not refer to aggression or competitive striving or any such mechanism. It is the individual affirming his existence and his potentialities as a being in his own right; it is "the courage to be as an individual," as Tillich remarks in his discussion of Nietzsche. The word *power* is used by Nietzsche in the classical sense of *potentia, dynamis.* Kaufmann succinctly summarizes Nietzsche's belief at this point:

> Man's task is simple: he should cease letting his "existence" be "a thoughtless accident." Not only the use of the word *Existenz,* but the thought which is at stake, suggests that [this essay] is particularly close to what is today called *Existenz-philosophie.* Man's fundamental problem is to achieve true "existence" instead of letting his life be no more than just another accident. In *The Gay Science* Nietzsche hits on a formulation which brings out the essential paradox of any distinction between self and true self: "What does your conscience say?—*You shall become who you are.*" Nietzsche maintains this

conception until the end, and the full title of his last work is *Ecce Homo, Wie man wird, was man ist*—how one becomes what one is.[13]

In an infinite variety of ways, Nietzsche holds that this power, this expansion, growing, bringing one's inner potentialities into birth in action is the central dynamic and need of life. His work here relates directly to the problem in psychology of what the fundamental drive of organisms is, the blocking of which leads to neurosis: it is not urge for pleasure or reduction of libidinal tension or equilibrium or adaptation. The fundamental drive rather is to live out one's *potentia.* "*Not* for pleasure does man strive," holds Nietzsche, "but for power."[14] Indeed, happiness is not absence of pain but "the most alive feeling of power,"[15] and joy is a "*plus-feeling* of power."[16] Health, also, he sees as a by-product of the use of power, power here specifically described as the ability to overcome disease and suffering.[17]

We come to Nietzsche's concept of being, the basic principle of his philosophy. Being is a "generalization," he writes, "of the concept of life, of willing, acting, and becoming." And then again,

> The soul in its essence will say to herself: no one can build the bridge on which you in particular will have to cross the river of life—no one but yourself. Of course there are countless paths and bridges and demigods ready to carry you over the river, but only at the price of your own self. In all the world, there is one specific way that no one but you can take. Whither does it lead? Do not ask, but walk it. As soon as one says, "I want to remain myself," he discovers that it is a frightful resolve. Now he must descend to the depths of his existence.

The tendency in the United States is to contrast *being* with *becoming.* The latter word is more acceptable to our American psychologists: Abraham Maslow used it continuously; Gordon Allport used it in the title of one of his books. Becoming is believed to represent the dynamic, the moving, the changing condition, in contrast to what is wrongly considered the ontological static quality of being. I believe this is an error. It is interesting that Nietzsche also, in his later life, comes to the same conclusion. He writes: "In becoming everything is hollow, illusory, flat; the riddle which man must solve can be solved only through being, a being which is just what it is, and cannot perish. Man is now beginning to gauge the depth of his fusion with becoming and with being."

Nietzsche was a naturalist in the sense that he sought at all times to relate every expression of life to the broad context of all of nature, but it is precisely at this point that he makes clear that human psychology is always more than biology. One of his most crucial existential emphases is his insistence that the values of *human* life never come about automatically. The human being *can lose his own being by his own choices,* as a tree or stone cannot. Affirming one's own being creates the values of life. "Individuality, worth and dignity are not *gegeben,* i.e., given us as data by nature, but *aufgegeben*—i.e., given or assigned to us as a task which we ourselves must solve."[18] This is an emphasis which likewise comes out in Tillich's belief that courage opens the way to being: if you do not have "courage to be," you lose your own being. And it similarly appears in extreme form in Sartre's contention, you *are* your choices.

In his approach to the question of health, Nietzsche again speaks to contemporary psychology. Health is not a fixed state that some lucky people arrive at. (This is a great consolation to many of us who have been ill much of our lives!) Health is a dynamic balance in the struggle to overcome disease. The artist is an artist because he gains a sensitivity out of his struggle between sickness and health. To quote Nietzsche here: "The spirit grows. Strength is restored by wounding." Health is the capacity to overcome disease. This points toward Nietzsche's later idea of "power" as the artist's ability to overcome disease and suffering.

We also find in Nietzsche continuous denial of the common idea that survival is the highest value of life. He heaps scorn on those who think they are Darwinians and who fail to see that man seeks not to preserve his potentiality, but rather to express it.

At almost any point at which one opens Nietzsche, one finds psychological insights which are not only penetrating and astute in themselves but amazingly parallel to the psychoanalytic mechanisms Freud was to formulate a decade and more later. For example, turning to the *Genealogy of Morals,* written in 1887, we find "All instincts that are not allowed free play turn inward. This is what I call man's *interiorization.*"[19] One looks twice, noting the curiously close prediction of the later Freudian concept of repression. Nietzsche's eternal theme was the unmasking of self-deception. Throughout the whole essay mentioned above he develops the thesis that altruism and morality are the results of repressed hostility and resentment, that when the individual's *potentia* are turned inward, bad conscience is the result. He gives a vivid description of the "impotent" people "who are full of bottled-up aggressions: their happiness is purely passive and takes the form of drugged tranquillity, stretching and yawning, peace, 'sabbath,' emotional slackness."[20] This in-turned aggression breaks out in sadistic demands on others—the process which later was to be designated in psychoanalysis as symptom formation. And the demands clothe themselves as morality—the process which Freud later called reaction formation. "In its earliest phase," Nietzsche writes, "bad conscience is nothing other than the instinct of freedom forced to become latent, driven underground, and forced to vent its energy upon itself." At other points we find staring us in the face striking formulations of sublimation, a concept which Nietzsche specifically developed. Speaking of the connection between a person's artistic energy and sexuality, he says that it "may well be that the emergence of the aesthetic condition does not suspend sensuality, as Schopenhauer believed, but merely *transmutes it in such a way that it is no longer experienced as a sexual incentive.*"[21]

Nietzsche asserts that joy does not come from submission and abnegation, but from assertion. "Joy is only a symptom," he writes, "of a feeling of attained power." The essence of joy, again, is a plus feeling of power. Far from being a destructive, nihilistic thinker, Nietzsche turns out, on deeper inspection, to be profoundly constructive. And he is constructive in a way that seems the only way for our day. In the degree of apathy —indeed the degree of neurotic apathy—that exists in our country, and the suppression, not only of anxiety, but even more deeply of guilt, we greatly need Nietzsche's gospel. This is why Nietzsche is the therapist for the therapists in our time.

FREUD AND NIETZSCHE

What, then, are we to conclude from this remarkable parallel between Nietzsche's ideas and Freud's? The similarity was known to the circle around Freud. One evening in 1908 the Vienna Psychoanalytic Society had as its program a discussion of Nietzsche's *Genealogy of Morals.* Freud mentioned that he had tried to read Nietzsche, but found Nietzsche's thought so rich he renounced the attempt. Freud then stated that "Nietzsche had a more penetrating knowledge of himself than any other man who ever lived or was ever likely to live."[22] This judgment, repeated on several occasions, was, as Jones remarks, no small compliment from the inventor of psychoanalysis. Freud always had a strong but ambivalent interest in philosophy; he distrusted and even feared it.[23] Jones points out that this distrust was on personal as well as intellectual grounds. One of the reasons was his suspicion of arid intellectual speculation—a point on which Kierkegaard, Nietzsche, and the other existentialists would have enthusiastically agreed with him. In any case, Freud felt that his own potential proclivity for philosophy "needed to be sternly checked, and for that purpose he chose the most effective agency—scientific discipline."[24] At another point Jones remarks, "The ultimate questions of philosophy were very near to him in spite of his endeavor to keep them at a distance and of distrusting his capacity to solve them."[25] Nietzsche's works may not have had a direct, but most certainly had an indirect, influence on Freud. It is clear that the ideas which were later to be formulated in psychoanalysis were "in the air" in the Europe of the end of the nineteenth century. The fact that Kierkegaard, Nietzsche, and Freud all dealt with the same problems of anxiety, despair, fragmentalized personality, and the symptoms of these bears out our earlier thesis that psychoanalysis and the existential approach to human crises were called forth by and were answers to the same problems. It does not detract from the genius of Freud to point out that probably almost all of the specific ideas which later appeared in psychoanalysis could be found in Nietzsche in greater breadth and in Kierkegaard in greater depth.

But the particular genius of Freud lies in his translating these depth-psychological insights into the natural scientific framework of his day. For this task he was admirably fitted—in temperament highly objective and rationally controlled, indefatigable, and capable of taking the infinite pains necessary for his systematic work. He did accomplish something new under the sun—namely, the transmuting into the scientific stream of Western culture the new psychological concepts, where they could be studied with some objectivity, built upon, and within certain limits rendered teachable.

But is not the very genius of Freud and of psychoanalysis likewise also its greatest danger and most serious shortcoming? For the translation of depth-psychological insights into objectivated science had results which could have been foreseen. One such result has been the limiting of the sphere of investigation in man to what fits this sphere of science. Binswanger points out that Freud deals only with the *homo natura* and that, whereas his methods admirably fitted him for exploring the *Umwelt,* the world of man in his biological environment, they by the same token prevented him from comprehending fully the *Mitwelt,* man in personal relations with fellow men, and the *Eigenwelt,* the sphere of man in relation to himself.[26] Another more serious practical result has been, as we shall indicate later in our discussion of the concepts of determinism and passivity of the ego, a new tendency to objectivate personality and to contribute to the very developments in modern culture which caused the difficulties in the first place.

We now come to a very important problem, and in order to understand it we need to make one more preliminary distinction. That is between *reason* as the term was used in the seventeenth century and the Enlightenment and *technical reason* today. Freud held a concept of reason which came directly from the Enlightenment—namely, "ecstatic reason." And he equated this with science. This use of reason involves, as seen in Spinoza and the other thinkers of the seventeenth and eighteenth centuries, a confidence that reason can by itself comprehend all problems. But those thinkers were using reason as including the capacity to transcend the immediate situation, to grasp the whole, and such functions as intuition, insight, poetic perception were not rigidly excluded. The concept also embraced ethics: reason in the Enlightenment meant justice. Much, in other words, that is called "irrational" in our day was included in their idea of reason. This accounts for the tremendous and enthusiastic faith they could lodge in it. But by the end of the nineteenth century, as Tillich demonstrates most cogently, this ecstatic character had been lost. Reason had become "technical reason": reason married to techniques, reason as functioning best when devoted to isolated problems, reason as an adjunct and subordinate to industrial progress, reason as separated from emotion and will, reason indeed as opposed to existence—the reason finally which Kierkegaard and Nietzsche so strongly attacked.

Now, part of the time Freud uses the concept of reason in the ecstatic form, as when he speaks of reason as "our salvation," reason as our "only recourse," and so on. Here one gets the anachronistic feeling that his sentences are directly out of Spinoza or some writer of the Enlightenment. Thus he tried on the one hand to preserve the ecstatic concept, tried to save the view of man and reason which transcends techniques. But, on the other hand, in equating reason with science, Freud makes it technical reason. His great contribution was his effort to overcome the fragmentation of man by bringing man's irrational tendencies into the light, bringing unconscious, split-off, and repressed aspects of personality into consciousness and acceptance. But the other side of his emphasis—namely, the identification of psychoanalysis with technical reason—is an expression of the precise fragmentation which he sought to cure. It is not unfair to say that the prevailing trend in the development of psychoanalysis in late decades, particularly after the death of Freud, has been to reject his efforts to save reason in its ecstatic form and to accept exclusively the latter—namely, reason in its technical form.

This trend is generally unnoticed since it fits in so well with dominant trends in our whole culture. But we have already noted that seeing man and his functions in their technical form is one of the central factors in the compartmentalization of contemporary man. Thus a critical and serious dilemma faces us. On the theoretical side, psychoanalysis (and other forms of psychology to the extent that they are wedded to technical reason) themselves add to the chaos in our theory of man, both scientific and philosophical, of which Cassirer and Scheler spoke. On the practical side, there is considerable danger that psychoanalysis, as well as other forms of psychotherapy and adjustment psychology, will become new representations of the fragmentation of man, that they will exemplify the loss of the individual's vitality and significance, rather than the reverse, that the new techniques will assist in standardizing and giving cultural sanction to man's alienation from himself rather than solving it, that they will become expressions of the new mechanization of man, now calculated and controlled with greater psychological precision and on the vaster scale of unconscious and depth dimensions—that psychoanalysis and psychotherapy in general will become part of the neurosis of our day rather than part of the cure. This would indeed be a supreme irony of history. It is not alarmism nor the showing of unseemly fervor to point out these tendencies, some of which are already upon us. It is simply to look directly at our historical situation and to draw unflinchingly the implications.

We are now in a position to see the crucial significance of the existential psychotherapy movement. *It is precisely the movement that protests against the tendency to identify psychotherapy with technical reason.* It stands for basing psychotherapy on an understanding of what makes man the *human* being; it stands for defining neurosis in terms of what destroys man's capacity to fulfill his own being. We have seen that Kierkegaard and Nietzsche, as well as the representatives of the existential cultural movement following them, not only contributed far-reaching and penetrating psychological insights, which in themselves form a significant contribution to anyone seeking scientifically to understand modern psychological problems, but also did something else—they placed these insights on an ontological basis, namely, the study of *man as the being who has these particular problems.* They believed that it was absolutely necessary that this be done, and they feared that the subordination of reason to technical problems would ultimately mean the making of man over in the image of the machine. Science, Nietzsche had warned, is becoming a factory, and the result will be ethical nihilism.

Existential psychotherapy is the movement which, although standing on one side on the scientific analysis owed chiefly to the genius of Freud, also brings back into the picture the understanding of man on the deeper and broader level—man as the being who is human. It is based on the assumption that it is possible to have a science of man which does not fragmentize man and destroy his humanity at the same moment it studies him. It unites science and ontology. It is not too much to say, thus, that we are here not merely discussing a new method as over against other methods, to be taken or to be left or to be absorbed into some vague catch-all eclecticism. The issues strike much deeper into our contemporary historical situation.

PART III

CONTRIBUTIONS TO THERAPY

SIX

To Be and Not to Be

THE FUNDAMENTAL CONTRIBUTION of existential therapy is its understanding of man as *being*. It does not deny the validity of dynamisms and the study of specific behavior patterns in their rightful places. But it holds that drives or dynamisms, by whatever name one calls them, can be understood only in the context of the structure of the existence of the person we are dealing with. The distinctive character of existential analysis is, thus, that it is concerned with *ontology*, the science of being, and with *Dasein*, the existence of this particular being sitting opposite the psychotherapist.

Before struggling with definitions of *being* and related terms, let us begin existentially by reminding ourselves that what we are talking about is an experience every sensitive therapist must have countless times a day. It is the experience of the instantaneous encounter with another person who comes alive to us on a very different level from what we know *about* him. "Instantaneous" refers not to the actual time involved but to the quality of the experience. We may know a great deal about a patient from his case record, let us say, and may have a fairly good idea of how other interviewers have described him. But when the patient himself steps in, we often have a sudden, sometimes powerful, experience of here is a new person, an experience that normally carries with it an element of surprise, not in the sense of perplexity or bewilderment, but in its etymological sense of being "taken from above." This is in no sense a criticism of one's colleagues' reports; for we have this experience of encounter even with persons we have known or worked with for a long time. We may have it with friends and loved ones. It is not a once-and-for-all experience; indeed, in any developing, growing relationship it may—probably should, if the relationship is vital—occur continually.

The data we learned *about* the patient may have been accurate and well worth learning. But the point, rather, is that *the grasping of the being of the other person occurs on a different level from our knowledge of specific things about him.* Obviously a knowledge of the drives and mechanisms which are in operation in the other person's behavior is useful; a familiarity with his patterns of interpersonal relationships is highly relevant; information about his social conditioning, the meaning of particular gestures and symbolic actions is to the point, and so on *ad infinitum.* But all these fall on to a different level when we confront the overarching, most real fact of all—namely, the immediate, living person himself. When we find that all our voluminous knowledge about the person suddenly forms itself into a new pattern in this confrontation, the implication is not that the knowledge was wrong; it is rather that it takes its meaning, form, and significance from the reality of the person of whom these specific things are expressions.

Nothing we are saying here in the slightest deprecates the importance of gathering and studying seriously all the specific data one can get about the given person. This is only common sense. But neither can one close his eyes to the experiential fact that these data form themselves into a configuration given in the encounter with the person himself. This also is illustrated by the common experience we all have had in interviewing persons; we may say we do not get a "feeling" of the other person and need to prolong the interview until the data "break" into their own form in our minds. We particularly do not get this "feeling" when we ourselves are hostile or resent the relationship—that is, keep the other person out—no matter how intellectually bright we may be at the time. This is the classical distinction between *knowing* and *knowing about.* When we seek to know a person, the knowledge *about* him must be subordinated to the overarching fact of his actual existence.

In the ancient Greek and Hebrew languages the verb *to know* is the same word as that which means "to have sexual intercourse." This is illustrated time and again in the King James translation of the Bible—"Abraham knew his wife and she conceived. . . ." *To know* had the same meaning in sixteenth- and seventeenth-century English. Thus the etymological relation between knowing and loving is exceedingly close. Knowing another human being, like loving him, involves a kind of union, a dialectical participation with the other. This Binswanger calls the "dual mode." One must have at least a readiness to love the other person, broadly speaking, if one is to be able to understand him.

The encounter with the being of another person has the power to shake one profoundly and may potentially be very anxiety-arousing. It may also be joy-creating. In either case, it has the power to grasp and move one deeply. The therapist understandably may be tempted for his own comfort to abstract himself from the encounter by thinking of the other as just a "patient" or by focusing only on certain mechanisms of behavior. But if the technical view is used dominantly in the relating to the other person, obviously one has defended oneself from anxiety at the price not only of the isolation of oneself from the other but also of radical distortion of reality. For one does not then really *see* the other person. It does not disparage the importance of technique to point out that technique, like data, must be subordinated to the fact of the reality of two persons in the room.

This point has been admirably made in a slightly different way by Sartre. If we "consider man," he writes, "as capable of being analyzed and reduced to original data, to determined drives (or 'desires'), supported by the subject as properties of an object," we may indeed end up with an imposing system of substances which we may then call mechanisms or dynamisms or patterns. But we find ourselves up against a dilemma. Our human being has become "a sort of indeterminate clay which would have to receive [the desires] passively—or he would be reduced to a simple bundle of these irreducible drives or tendencies. In either case the *man* disappears; we can no longer find 'the one' to whom this or that experience has happened."[1]

It is difficult enough to give definitions of *being* and *Dasein,* but our task is made doubly difficult by the fact that these terms and their connotations encounter much resistance. Some readers may feel that these words are only a new form of "mysticism" (used in its disparaging and quite inaccurate sense of "misty") and have nothing to do with science. But this attitude obviously dodges the whole issue by disparaging it. It is interesting that the term *mystic* is used in this derogatory sense to mean anything we cannot segmentize and count. The odd belief prevails in our culture that a thing or experience is not real if we cannot make it mathematical, and somehow it must be real if we can reduce it to numbers. But this means making an abstraction out of it—mathematics is the abstraction par excellence, which is indeed its glory and the reason for its great usefulness. Modern Western man thus finds himself in the strange situation, after reducing something to an abstraction, of having then to persuade himself it is real. This has much to do with the sense of isolation and loneliness which is endemic in the modern Western world; for the only experience we let ourselves believe in as real is that which precisely is not. Thus we deny the reality of our own experience. The term *mystic,* in this disparaging sense, is generally used in the service of obscurantism; certainly avoiding an issue by derogation is only to obscure it. Is not the scientific attitude, rather, to try to see clearly what it is we are talking about and then to find whatever terms or symbols can best, with least distortion, describe this reality? It should not so greatly surprise us to find that "being" belongs to that class of realities, like "love" and "consciousness" (for two other examples), which we cannot segmentize or abstract without losing precisely what we set out to study. This does not, however, relieve us from the task of trying to understand and describe them.

A more serious source of resistance is one that runs through the whole of modern Western society—namely, the psychological need to avoid and, in some ways, repress, the whole concern with "being." In contrast to other cultures which may be very concerned with being—particularly Indian and East Asian—and other historical periods which have been so concerned, the characteristic of our period in the West, as Marcel rightly phrases it, is precisely that the awareness of "the sense of the ontological—the sense of being—is lacking. Generally speaking, modern man is in this condition; if ontological demands worry him at all, it is only dully, as an obscure impulse."[2] Marcel points out what many students have emphasized, that this loss of the sense of being is related on one hand to our tendency to subordinate existence to function: a man knows himself not as a man or self but as a token seller in the subway, a grocer, a professor, a vice-president of AT&T, or by whatever his economic function may be. And on the other hand, this loss of the sense of being is related to the mass collectivist trends and widespread conformist tendencies in our culture. Marcel then makes this trenchant challenge: "*Indeed I wonder if a psychoanalytic method, deeper and more discerning than any that has been evolved until now, would not reveal the morbid effects of the repression of this sense and of the ignoring of this need.*"[3]

"As for defining the word 'being,' " Marcel goes on, "let us admit that it is extremely difficult; I would merely suggest this method of approach: being is what withstands—or what would withstand—an exhaustive analysis bearing on the data of experience and aiming to reduce them step by step to elements increasingly devoid of intrinsic or significant value. (An analysis of this kind is attempted in the theoretical works of Freud.)"[4] This last sentence I take to mean that when Freud's analysis is pushed to the ultimate extreme, and we know, let us say, everything about drives, instincts, and mechanisms, we have everything *except* being. Being is that which remains. It is that which constitutes this infinitely complex set of deterministic factors into a person *to whom* the experiences happen and who possesses some element, no matter how minute, of freedom to become aware that these forces are acting upon him. This is the sphere where he has the potential capacity to pause before reacting and thus to cast some weight on whether his reaction will go this way or that. And this, therefore, is the sphere where he, the human being, is never merely a collection of drives and determined forms of behavior.

The term the existential therapists use for the distinctive character of human existence is *Dasein.* Binswanger, Kuhn, and others designate their schools as *Daseinsanalyse.* Composed of *sein* (being) plus *da* (there), *Dasein* indicates that man is the being who *is there* and implies also that he *has* a "there" in the sense that he can know he is there and can take a stand with reference to that fact. The "there" is moreover not just any place, but the particular "there" that is mine, the particular point *in time* as well as space of my existence at this given moment. Man is the being who can be conscious of, and therefore responsible for, his existence. It is this capacity to become aware of his own being which distinguishes the human being from other beings. The existential therapists think of man not only as "being in itself," as all beings are, but also as "being for itself." Binswanger speaks of "*Dasein* choosing" this or that, meaning "the person-who-is-responsible-for-his-existence choosing."

The full meaning of the term *human being* will be clearer if the reader will keep in mind that *being* is a participle, a verb form implying that someone is in the process of *being something.* It is unfortunate that, when used as a general noun in English, the term *being* connotes a static substance, and when used as a particular noun such as *a* being, it is usually assumed to refer to an entity, say, such as a soldier to be counted as a unit. Rather, *being* should be understood, when used as a general noun, to mean *potentia,* the source of potentiality; *being* is the potentiality by which the acorn becomes the oak or each of us becomes what he truly is. And when used in a particular sense, such as *a* human being, it always has the dynamic connotation of someone in process, the person being something. Perhaps, therefore, *becoming* connotes more accurately the meaning of the term in this country, despite the difficulties with the term we have mentioned earlier. We can understand another human being only as we see what he is moving toward, what he is becoming; and we can know ourselves only as we "project our *potentia* in action." The significant tense for human beings is thus the *future*—that is to say, the critical question is what I am pointing toward, what I will be in the immediate future.

Thus, being in the human sense is not given once and for all. It does not unfold automatically as the oak tree does from the acorn. For an intrinsic and inseparable element in being human is self-consciousness. Man (or *Dasein*) is the particular being who has to be aware of himself, be responsible for himself, if he is to become himself. He also is that particular being who knows that at some future moment he will not be; he is the being who is always in a dialectical relation with nonbeing, death. And he not only knows he will sometime not be, but he can, in his own choices, slough off and forfeit his being. "To be and not to be"—the "and" in our title to this chapter is not a typographical error—is not a choice one makes once and for all at the point of considering suicide; it reflects to some degree a choice made at every instant. The profound dialectic in the human being's awareness of his own being is pictured with incomparable beauty by Pascal:

> Man is only a reed, the feeblest reed in nature, but he is a thinking reed. There is no need for the entire universe to arm itself in order to annihilate him: a vapour, a drop of water, suffices to kill him. But were the universe to crush him, man would yet be more noble than that which slays him, because he knows that he dies, and the advantage that the universe has over him; of this the universe knows nothing.[5]

In the hope of making clearer what it means for a person to experience his own being, we shall present an illustration from a case history. This patient, an intelligent woman of twenty-eight, was especially gifted in expressing what was occurring within her. She had come for psychotherapy because of serious anxiety spells in closed places, severe self-doubts, and eruptions of rage which were sometimes uncontrollable.[6] An illegitimate child, she had been brought up by relatives in a small village in the southwestern part of the country. Her mother, in periods of anger, often reminded her as a child of her origin, recounted how she had tried to abort her, and in times of trouble had shouted at the little girl, "If you hadn't been born, we wouldn't have to go through this!" Other relatives had cried at the child, in family quarrels, "Why didn't you kill yourself?" and "You should have been choked the day you were born!" Later, as a young woman, the patient had become well educated on her own initiative.

In the fourth month of therapy she had the following dream: "I was in a crowd of people. They had no faces; they were like shadows. It seemed like a wilderness of people. Then I saw there was someone in the crowd who had compassion for me." The next session she reported that she had had, in the intervening day, an exceedingly important experience. It is reported here as she wrote it down from memory and notes two years later.

I remember walking that day under the elevated tracks in a slum area, feeling the thought, "I am an illegitimate child." I recall the sweat pouring forth in my anguish in trying to accept that fact. Then I understood what it must feel like to accept, "I am a Negro in the midst of privileged whites," or "I am blind in the midst of people who see." Later on that night I woke up and it came to me this way, "I accept the fact that I am an illegitimate child." *But* "I am not a child anymore." So it is, "I am illegitimate." That is not so either: "I was born illegitimate." Then what is left? What is left is this, *"I Am."* This act of contact and acceptance with "I am," once gotten hold of, gave me (what I think was for me the first time) the experience "Since I Am, I have the right to be."

What is this experience like? It is a primary feeling—it feels like receiving the deed to my house. It is the experience of my own aliveness not caring whether it turns out to be an ion or just a wave. It is like when a very young child I once reached the core of a peach and cracked the pit, not knowing what I would find and then feeling the wonder of finding the inner seed, good to eat in its bitter sweetness. . . . It is like a sailboat in the harbor being given an anchor so that, being made out of earthly things, it can by means of its anchor get in touch again with the earth, the ground from which its wood grew; it can lift its anchor to sail but always at times it can cast its anchor to weather the storm or rest a little. . . . It is my saying to Descartes, *"I Am, therefore* I think, I feel, I do."

It is like an axiom in geometry—never experiencing it would be like going through a geometry course not knowing the first axiom. It is like going into my very own Garden of Eden where I am beyond good and evil and all other human concepts. It is like the experience of the poets of the intuitive world, the mystics, except that instead of the pure feeling of and union with God it is the finding of and the union with my own being. It is like owning Cinderella's shoe and looking all over the world for the foot it will fit and realizing all of a sudden that one's own foot is the only one it will fit. It is a "Matter of Fact" in the etymological sense of the expression. It is like a globe before the mountains and oceans and continents have been drawn on it. It is like a child in grammar finding the *subject* of the verb in a sentence—in this case the subject being one's own life span. It is ceasing to feel like a theory toward one's self. . . .

We shall call this the "I am" experience.[7] This one phase of a complex case, powerfully and beautifully described above, illustrates the emergence and strengthening of the sense of being in one person. The experience is etched the more sharply in this person because of the more patent threat to her being that she had suffered as an illegitimate child and her poetic articulateness as she looked back on her experience from the vantage point of two years later. I do not believe either of these facts, however, makes her experience different in fundamental quality from what human beings in general, normal or neurotic, go through.

We shall make four final comments on the experience exemplified in this case. First, the "I am" experience is not in itself the solution to a person's problems; it is rather the *precondition* for their solution. This patient spent some two years thereafter working through specific psychological problems, which she was able to do on the basis of this emerged experience of her own existence. In the broadest sense, the achieving of the sense of being is a goal of all therapy, but in the more precise sense it is a relation to oneself and one's world, an experience of one's own existence (including one's own identity), which is a prerequisite for the working through of specific problems. It is, as the patient wrote, the "primary fact," a *ur* experience. It is not to be identified with any patient's discovery of his specific powers—when he learns, let us say, that he can paint or write or work successfully or have successful sexual intercourse. Viewed from the outside, the discovery of specifice powers and the experience of one's own being may seem to go hand in hand, but the latter is the underpinning, the foundation, the psychological precondition of the former.

We may well be suspicious that solutions to a person's specific problems in psychotherapy which do not presuppose this "I am" experience in greater or lesser degree will have a pseudo quality. The new "powers" the patient discovers may well be experienced by him as merely compensatory—that is, as proofs that he is of significance despite the fact that he is certain on a deeper level that he is not, since he still lacks a basic conviction of "*I Am,* therefore I think, I act." And we could well wonder whether such compensatory solutions would not represent rather the patient's simply exchanging one defense system for another, one set of terms for another, without ever experiencing himself as alive, significant, and existing. In the second state the patient, instead of blowing up in anger, "sublimates" or "introverts" or "relates," but still without the act being rooted in his own existence.

Our second comment is that this patient's "I am" experience is not to be explained by the transference relationship. That the positive transference, whether directed to therapist or husband,[8] is obviously present in the above case is shown in the eloquent dream the night before in which there was one person in the depersonalized wilderness of the crowd—I assume that this is myself, the therapist—who had compassion for her. True, she is showing in the dream that she could have the "I am" experience only if she could trust some other human being. But this does not account for the experience itself. It may well be true that for any human being the possibility of acceptance by and trust for another human being is a necessary condition for the "I am" experience. But the awareness of one's own being occurs basically on the level of the grasping of oneself; it is an experience of *Dasein,* realized in the realm of self-awareness. It is not to be explained *essentially* in social categories. The acceptance by another person, such as the therapist, shows the patient that he no longer needs to fight his main battle on the front of whether anyone else, or the world, can accept him; the acceptance *frees* him to experience his own being. This point must be emphasized because of the common error in many circles of assuming that the experience of one's own being will be discovered automatically if only one is accepted by somebody else. This is the basic error of some forms of "relationship therapy." The attitude of "If I love and accept you, this is all you need" is in life and in therapy an attitude which may well minister to increased passivity. The crucial question is what the individual himself, in his own awareness of and responsibility for his existence, does with the fact that he can be accepted.

The third comment follows directly from the above, that *being* is a category which cannot be reduced to introjection of social and ethical norms. It is, to use Nietzsche's phrase, "beyond good and evil." To the extent that my sense of existence is authentic, it is precisely *not* what others have told me I should be, but is the one Archimedes point I have to stand on from which to judge what parents and other authorities demand. Indeed, *compulsive and rigid moralism arises in given persons precisely as the result of a lack of a sense of being.* Rigid moralism is a compensatory mechanism by which the individual persuades himself to take over the external sanctions because he has no fundamental assurance that his own choices have any sanction of their own. This is not to deny the vast social influences in anyone's morality, but it is to say that the ontological sense cannot be wholly reduced to such influences. The ontological sense *is not a superego* phenomenon. By the same token the sense of being gives the person a basis for a self-esteem which is not merely the reflection of others' views about him. For if your self-esteem must rest in the long run on social validation, you have not self-esteem but a more sophisticated form of social conformity. It cannot be said too strongly that the sense of one's own existence, though interwoven with all kinds of social relatedness, is in basis not the product of social forces; it always presupposes *Eigenwelt,* the "own world" (a term which will be discussed below).

Our fourth comment deals with the most important consideration of all—namely, that the "I am" experience must not be identified with what is called in various circles the "functioning of the ego." That is to say, it is an error to define the emergence of awareness of one's own being as one phase of the "development of the ego." We need only reflect on what the concept of "ego" has meant in classical psychoanalytic tradition to see why this is so. The ego was traditionally conceived as a relatively weak, shadowy, passive, and derived agent, largely an epiphenomenon of other more powerful processes. It is "derived from the Id by modifications imposed on it from the external world" and is "representative of the external world."[9] "What we call the ego is essentially passive," says Groddeck, a statement which Freud cites with approval.[10] The developments in the middle period of psychoanalytic theory brought increased emphasis on the ego, to be sure, but chiefly as an aspect of the study of defense mechanisms; the ego enlarged its originally buffeted and frail realm chiefly by its negative defensive functions. It "owes service to three masters and is consequently menaced by three dangers: the external world, the libido of the Id, the severity of the Super-ego."[11] Freud often remarked that the ego does very well indeed if it can preserve some semblance of harmony in its unruly house.

A moment's thought will show how great is the difference between this ego and the "I am" experience, the sense of being which we have been discussing. The latter occurs on a more fundamental level and is a precondition for ego development. The ego is a *part* of the personality, and traditionally a relatively weak part, whereas the sense of being refers to one's whole experience, unconscious as well as conscious, and is by no means merely the agent of awareness. The ego is a reflection of the outside world; the sense of being is rooted in one's own experience of existence, and if it is a mirroring of, a reflection of, the outside world alone, it is then precisely not one's own sense of existence. My sense of being is *not* my capacity to see the outside world, to size it up, to assess reality; it is rather my capacity to see myself as a being in the world, *to know myself as the being who can do these things.* It is in this sense a precondition for what is called "ego development." The ego is the *subject* in the subject-object relationship; the sense of being occurs on a level prior to this dichotomy. Being means not "I am the subject," but "I am the being who can, among other things, know myself as the subject of what is occurring." The sense of being is not in origin set against the outside world but it must include this capacity to set oneself against the external world if necessary, just as it must include the capacity to confront nonbeing. To be sure, both what is called the ego and the sense of being presuppose the emergence of self-awareness in the child somewhere between the first couple of months of infancy and the age of two years, a developmental process often called the "emergence of the ego." But this does not mean these two should be identified. The ego is said normally to be especially weak in childhood, weak in proportion to the child's relatively weak assessment of and relation to reality; whereas the sense of being may be especially strong, only later to diminish as the child learns to give himself over to conformist tendencies, to experience his existence as a reflection of others' evaluation of him, to lose some of his originality and primary sense of being. Actually, the sense of being—that is, the ontological sense—is presupposed for ego development, just as it is presupposed for the solution of other problems.[12]

We are aware that additions and elaborations are occurring in ego theory of late decades in the orthodox psychoanalytic tradition. But one cannot strengthen such a weak monarch by decking him out in additional robes, no matter how well-woven or intricately tailored the robes may be. The real and fundamental trouble with the doctrine of the ego is that it represents, par excellence, the subject-object dichotomy in modern thought. Indeed, it is necessary to emphasize that *the very fact that the ego is conceived of as weak, passive, and derived is itself an evidence and a symptom of the loss of the sense of being in our day, a symptom of the repression of the ontological concern.* This view of the ego is a symbol of the pervasive tendency to see the human being primarily as a passive recipient of forces acting upon him, whether the forces be identified as the Id or the vast industrial juggernaut in Marxian terms or the submersion of the individual as "one among many" in the sea of conformity, in Heidegger's terms. The view of the ego as relatively weak and buffeted about by the Id was in Freud a profound symbol of the fragmentation of man in the Victorian period and also a strong corrective to the superficial voluntarism of that day. But the error arises when this ego is elaborated as the basic norm. The sense of being, the ontological awareness, must be assumed below ego theory if that theory is to refer with self-consistency to the human being as human.

We now come to the important problem of *nonbeing* or, as phrased in existential literature, *nothingness.* The "and" in the title of this chapter, "To Be *and* Not to Be," expresses the fact that nonbeing is an inseparable part of being. To grasp what it means to exist, one needs to grasp the fact that he might not exist, that he treads at every moment on the sharp edge of possible annihilation and can never escape the fact that death will arrive at some unknown moment in the future. Existence, never automatic, not only can be sloughed off and forfeited but is indeed at every instant threatened by nonbeing. Without this awareness of nonbeing—that is, awareness of the threats to one's being in death, anxiety, and the less dramatic but persistent threats of loss of potentialities in conformism—existence is vapid, unreal, and characterized by lack of concrete self-awareness. But with the confronting of nonbeing, existence takes on vitality and immediacy, and the individual experiences a heightened consciousness of himself, his world, and others around him.

Death is the most obvious form of the threat of nonbeing.

Freud grasped this truth on one level in his symbol of the death instinct. Life forces (being) are arrayed at every moment, he held, against the forces of death (nonbeing), and in every individual life the latter will ultimately triumph. But Freud's concept of the death instinct is an ontological truth and should not be taken as a deteriorated psychological theory. The concept of the death instinct is an excellent example of the point that Freud went beyond technical reason and tried to keep open the tragic dimension of life. His emphasis on the inevitability of hostility, aggression, and self-destructiveness in existence also, from one standpoint, has this meaning. True, he phrased these concepts wrong, as when he interpreted the "death instinct" in chemical terms. The use of the word *thanatos* in psychoanalytic circles as parallel to libido is an example of this deteriorated phraseology. These are errors which arise from trying to put ontological truths, which death and tragedy are, into the frame of technical reason and reduce them to specific psychological mechanisms. On that basis Horney and others could logically argue that Freud was too "pessimistic" and that he merely rationalized war and aggression. I think that is a sound argument against the usual oversimplified psychoanalytic interpretations, which are in the form of technical reason; but it is not a sound argument against Freud himself, who tried to preserve a real concept of tragedy, ambivalent though his frame of reference was. He had indeed a sense of nonbeing, despite the fact that he always tried to subordinate it and his concept of being to technical reason.

It is also an error to see the "death instinct" only in biological terms, which would leave us hobbled with a fatalism. The unique and crucial fact, rather, is that the human being is the one who *knows* he is going to die, who anticipates his own death. The critical question thus is how he relates to the fact of death: whether he spends his existence running away from death or making a cult of repressing the recognition of death under the rationalizations of beliefs in automatic progress or providence, as is the habit of our Western society, or obscuring it by saying "one dies" and turning it into a matter of public statistics which serve to cover over the one ultimately important fact, that he himself at some unknown future moment will die.

The existential analysts, on the other hand, hold that the confronting of death gives the most positive reality to life itself. It makes the individual existence real, absolute, and concrete. For "death as an irrelative potentiality singles man out and, as it were, individualizes him to make him understand the potentiality of being in others [as well as in himself], when he realizes the inescapable nature of his own death."[13] Death is, in other words, the one fact of my life which is not relative but absolute, and my awareness of this gives my existence and what I do each hour an absolute quality.

Nor do we need to go as far as the extreme example of death to discover the problem of nonbeing. Perhaps the most ubiquitous and ever-present form of the failure to confront nonbeing in our day is in *conformism,* the tendency of the individual to let himself be absorbed in the sea of collective responses and attitudes, to become swallowed up in *das Mann,* with the corresponding loss of his own awareness, potentialities, and whatever characterizes him as a unique and original being. The individual temporarily escapes the anxiety of nonbeing by this means, but at the price of forfeiting his own powers and sense of existence.

On the positive side, the capacity to confront nonbeing is illustrated in one's ability to accept anxiety, hostility, and aggression. By "accept" we mean here to tolerate without repression and so far as possible to utilize constructively. Severe anxiety, hostility, and aggression are states and ways of relating to oneself and others which would curtail or destroy being. But to preserve one's existence by running away from situations which would produce anxiety or situations of potential hostility and aggression leaves one with the vapid, weak, unreal sense of being—what Nietzsche meant in his brilliant description of the "impotent people" who evade their aggression by repressing it and thereupon experience "drugged tranquillity" and free-floating resentment. Our point does not at all imply the sloughing over of the distinction between the *neurotic* and *normal* forms of anxiety, hostility, and aggression. Obviously the one constructive way to confront neurotic anxiety, hostility, and aggression is to clarify them psychotherapeutically and so far as possible to wipe them out. But that task has been made doubly difficult, and the whole problem confused, by our failure to see the normal forms of these states—"normal" in the sense that they inhere in the threat of nonbeing with which all beings have to cope. Indeed, is it not clear that *neurotic* forms of anxiety, hostility, and aggression develop precisely because the individual has been unable to accept and deal with the *normal* forms of these states and ways of behaving? Paul Tillich has suggested far-reaching implications for the therapeutic process in his powerful sentence, which we shall quote without attempting to elucidate, "The self-affirmation of a being is the stronger the more nonbeing it can take into itself."

SEVEN

Anxiety and Guilt as Ontological

OUR DISCUSSION of being and nonbeing now leads to the point where we can understand the fundamental nature of anxiety. Anxiety is not an affect among other affects, such as pleasure or sadness. It is rather an ontological characteristic of man, rooted in his very existence as such. It is not a peripheral threat which I can take or leave, for example, or a reaction which may be classified beside other reactions; it is always a threat to the foundation, the center of my existence. Anxiety is *the experience of the threat of imminent nonbeing.*[1]

In his classical contributions to the understanding of anxiety, Kurt Goldstein has emphasized that anxiety is not something we "have" but something we "are." His vivid descriptions of anxiety at the onset of psychosis, when the patient is literally experiencing the threat of dissolution of the self, make his point abundantly clear. But, as he himself insists, this threat of dissolution of the self is not merely something confined to psychotics but describes the neurotic and normal nature of anxiety as well. Anxiety is the subjective state of the individual's becoming aware that his existence can become destroyed, that he can lose himself and his world, that he can become "nothing."[2]

This understanding of anxiety as ontological illuminates the difference between anxiety and fear. The distinction is not one of degree nor of the intensity of the experience. The anxiety a person feels when someone he respects passes him on the street without speaking, for example, is not as intense as the fear he experiences when the dentist seizes the drill to attack a sensitive tooth. But the gnawing threat of the slight on the street may hound him all day long and torment his dreams at night, whereas the feeling of fear, though it was quantitatively greater, is gone for the time being as soon as he steps out of the dentist's chair. The difference is that the anxiety strikes at the central core of his self-esteem and his sense of value as a self, which is the most important aspect of his experience of himself as a being. Fear, in contrast, is a threat to the periphery of his existence; it can be objectivated, and the person can stand outside and look at it. In greater or lesser degree, anxiety overwhelms the person's discovery of being, blots out the sense of time, dulls the memory of the past, and erases the future[3] —which is perhaps the most compelling proof of the fact that it attacks the center of one's being. While we are subject to anxiety, we are to that extent unable to conceive in imagination how existence would be "outside" the anxiety. This is why anxiety is so hard to bear, and why people will choose, if they have the chance, severe physical pain which would appear to the outside observer much worse. Anxiety is ontological, fear is not. Fear can be studied as an affect among other affects, a reaction among other reactions. But anxiety can be understood only as a threat to being itself.

This understanding of anxiety as an ontological characteristic again highlights our difficulty with words. The term which Freud, Binswanger, Goldstein, Kierkegaard (as he is translated into German) use for anxiety is *Angst,* a word for which there is no English equivalent. It is first cousin to *anguish* (which comes from Latin *angustus*, "narrow," which in turn comes from *angere*, "to pain by pushing together," "to choke"). The English term *anxiety*, such as in "I am anxious to do this or that," is a much weaker word.[4] Hence some students translate *Angst* as "dread," as did Lowrie in his now outdated translations of Kierkegaard. Some of us have tried to preserve the term *anxiety* for *Angst*[5] but we were caught in a dilemma. It seemed the alternative was either to use *anxiety* as a watered-down affect among other affects, which will work scientifically but at the price of the loss of power of the word; or to use such a term as *dread*, which carries literary power but has no role as a scientific category. Hence so often laboratory experiments on anxiety have seemed to fall woefully short of dealing with the power and devastating qualities of anxiety which we observe every day in clinical work, and also even clinical discussions about neurotic symptoms and psychotic conditions seem often to coast along the surface of the problem. The upshot of the existential understanding of anxiety is to give the term back its original power. It is an experience of threat which carries both anguish and dread, indeed the most painful and basic threat which any being can suffer, for it is the threat of loss of being itself. In my judgment, our psychological and psychiatric dealings with anxiety phenomena of all sorts will be greatly helped by shifting the concept to its ontological base.

Another significant aspect of anxiety may now also be seen more clearly—namely, the fact that anxiety always involves inner conflict. Is not this conflict precisely between what we have called being and nonbeing? Anxiety occurs at the point where some emerging potentiality or possibility faces the individual, some possibility of fulfilling his existence; but this very possibility involves the destroying of present security, which thereupon gives rise to the tendency to deny the new potentiality. Here lies the truth of the symbol of the birth trauma as the prototype of all anxiety—an interpretation suggested by the etymological source of the word *anxiety* as "pain in narrows," "choking," as though through the straits of being born. This interpretation of anxiety as birth trauma was, as is well known, held by Rank to cover all anxiety and was agreed to by Freud on a less comprehensive basis. There is no doubt that it carries an important symbolic truth even if one does not take it as connected with the literal birth of the infant. If there were not some possibility of opening up, some potentiality crying to be "born," we would not experience anxiety. This is why anxiety is so profoundly connected with the problem of freedom. If the individual did not have some freedom, no matter how minute, to fulfill some new potentiality, he would not experience anxiety. Kierkegaard described anxiety as "the dizziness of freedom," and added more explicitly, if not more clearly, "Anxiety is the reality of freedom as a potentiality before this freedom has materialized." Goldstein illustrates this by pointing out how people individually and collectively surrender freedom in the hope of getting rid of unbearable anxiety, citing the individual's retreating behind the rigid stockade of dogma or whole groups collectively turning to fascism in the interwar years in Europe.[6] In whatever way one chooses to illustrate it, this discussion points to the positive aspect of *Angst.* For the experience of anxiety itself demonstrates that some potentiality is present, some new possibility of being, threatened by nonbeing.

We have stated that the condition of the individual when confronted with the issue of fulfilling his potentialities is *anxiety.* We now move on to state that when the person denies these potentialities, fails to fulfill them, his condition is *guilt.* That is to say, guilt is also an ontological characteristic of human existence.[7]

This can be no better illustrated than to summarize a case Medard Boss cites of a severe obsessional-compulsive which he treated.[8] This patient, a physician suffering from washing, cleaning compulsions, had gone through both Freudian and Jungian analyses. He had had for some time a recurrent dream involving church steeples which had been interpreted in the Freudian analysis in terms of phallic symbols and in the Jungian in terms of religious archetype symbols. The patient could discuss these interpretations intelligently and at length, but his neurotic compulsive behavior, after temporary abeyance, continued as crippling as ever. During the first months of his analysis with Boss, the patient reported a recurrent dream in which he would approach a lavatory door which would always be locked. Boss confined himself to asking each time only why the door needed to be locked—to "rattling the doorknob," as he put it. Finally the patient had a dream in which he went through the door and found himself inside a church, waist deep in feces and being tugged by a rope wrapped around his waist leading up to the bell tower. The patient was suspended in such tension that he thought he would be pulled to pieces. He then went through a psychotic episode of four days during which Boss remained by his bedside, after which the analysis continued with an eventual very successful outcome.

Boss points out in his discussion of this case that the patient was guilty because he had locked up some essential potentialities in himself. *Therefore* he had guilt feelings. If, as Boss puts it, we "forget being"—by failing to bring ourselves to our entire being, by failing to be authentic, by slipping into the conformist anonymity of *das Mann*— then we have in fact missed our being and to that extent are failures. "If you lock up potentialities, you are guilty against (or *indebted to,* as the German word may be translated) what is given you in your origin, in your 'core.' In this existential condition of being indebted and being guilty are founded all guilt feelings, in whatever thousand and one concrete forms and malformations they may appear in actuality." This is what had happened to the patient. He had locked up both the bodily and the spiritual possibilities of experience (the "drive" aspect and the "god" aspect, as Boss also phrases it). The patient had previously accepted the libido and archetype explanations and knew them all too well; but that is a good way, says Boss, to escape the whole thing. Because the patient did not accept and take into his existence these two aspects, he was guilty, indebted to himself. This was the origin *(Anlass)* of his neurosis and psychosis.

The patient, in a letter to Boss sometime after the treatment, pointed out that the reason he could not really accept his anality in his first analysis was that he "sensed the ground was not fully developed in the analyst himself." The analyst had always attempted to reduce the dream of the church steeple to genital symbols and the "whole weight of the holy appeared to him as a mere sublimation mist." By the same token, the archetypal explanation, also symbolic, never could be integrated with the bodily, and for that matter never did really mesh with the religious experience either.

Let us note well that Boss says the patient *is* guilty, not merely that he *has guilt feelings.* This is a radical statement with far-reaching implications. It is an existential approach which cuts through the dense fog which has obscured much of the psychological discussion of guilt—discussions that have proceeded on the assumption that we can deal only with some vague "guilt feelings," as though it did not matter whether guilt was real or not. Has not this reduction of guilt to mere guilt feelings contributed considerably to the lack of reality and the sense of illusion in much psychotherapy? Has it not also tended to confirm the patient's neurosis in that it implicitly opens the way for him not to take his guilt seriously and to make peace with the fact that he has indeed forfeited his own being? Boss's approach is radically existential in that it takes the real phenomena with respect, here the real phenomenon being guilt. Nor is the guilt exclusively linked up with the religious aspect of this, or any patient's, experience: we can be as guilty by refusing to accept the anal, genital, or any other corporeal aspects of life as the intellectual or spiritual aspects. This understanding of guilt has nothing whatever to do with a judgmental attitude toward the patient. It has only to do with taking the patient's life and experience seriously and with respect.

We have cited only one form of ontological guilt—namely, that arising from forfeiting one's own potentialities. There are other forms as well. Another, for example, is ontological guilt against one's fellows, arising from the fact that since each of us is an individual, each necessarily perceives his fellow man through his own limited and biased eyes. This means that he always to some extent does violence to the true picture of his fellow man and always to some extent fails fully to understand and meet the other's needs. This is not a question of moral failure or slackness—though it can indeed be greatly increased by lack of moral sensitivity. It is an inescapable result of the fact that each of us is a separate individuality and has no choice but to look at the world through his own eyes. This guilt, rooted in our existential structure, is one of the most potent sources of a sound humility and an unsentimental attitude of forgiveness toward one's fellow men.

The first form of ontological guilt mentioned above—namely, forfeiting of potentialities—corresponds roughly to the mode of world which we shall describe and define in Chapter 9 called *Eigenwelt,* or own world. The second form of guilt corresponds roughly to *Mitwelt,* since it is guilt chiefly related to one's fellow men. (*Mitwelt* and *Eigenwelt* are also to be explained later.) There is a third form of ontological guilt which involves *Umwelt* as well as the other two modes—namely, "separation guilt" in relation to nature as a whole. This is the most complex and comprehensive aspect of ontological guilt. It may seem confusing, particularly since we are unable in this outline to explicate it in detail; we include it for the sake of completeness and for the interest of those who may wish to do further research in areas of ontological guilt. This guilt with respect to our separation from nature may well be much more influential (though repressed) than we realize in our modern

Western scientific age. It was originally expressed beautifully in a classical fragment from one of the early Greek philosophers of being, Anaximander: "The source of things is the boundless. From whence they arise, thence they must also of necessity return. For they do penance and make compensation to one another for their injustice in the order of time."

Ontological guilt has, among others, these characteristics. First, everyone participates in it. No one of us fails to some extent to distort the reality of his fellow men, and no one fully fulfills his own potentialities. Each of us is always in a dialectical relation to his potentialities, dramatically illustrated in the dream of Boss's patient being stretched between feces and bell tower. Second, ontological guilt does not come from cultural prohibitions, or from introjection of cultural mores; it is rooted in the fact of self-awareness. Ontological guilt does not consist of I am guilty because I violate parental prohibitions, but arises from the fact that I can see myself as the one who can choose or fail to choose. Every developed human being would have this ontological guilt, though its *content* would vary from culture to culture and would largely be given by the culture.

Third, ontological guilt is not to be confused with morbid or neurotic guilt. If it is unaccepted and repressed, it may turn into neurotic guilt. Just as neurotic anxiety is the end product of unfaced normal ontological anxiety, so neurotic guilt is the result of unconfronted ontological guilt. If the person can become aware of it and accept it (as Boss's patient later did), it is not morbid or neurotic. Fourth, ontological guilt does not lead to symptom formation, but has constructive effects in the personality. Specifically, it can and should lead to humility, sensitivity in one's relationships with one's fellow human beings, and increased creativity in the use of one's own potentialities.

EIGHT

Being in the World

SECOND ONLY in importance to the existential therapists' search for being is their concern with the person in his world. "To understand the compulsive," writes Erwin Straus, "we must first understand his world." And this is certainly true of all other types of patients as well as any human being, for that matter. For being together means *being together in the same world;* and knowing means knowing in the context of the same world. The world of this particular patient must be grasped from the inside, be known and seen so far as possible from the angle of the one who exists in it. "We psychiatrists," writes Binswanger, "have paid far too much attention to the deviations of our patients from life in the world which is common to all, instead of focusing primarily upon the patients' own or private world, as was first systematically done by Freud."[1]

The problem is how we are to understand the other person's world. It cannot be understood as an external collection of objects which we view from the outside (in which case we never really understand it), nor by sentimental identification (in which case our understanding doesn't do any good, for we have failed to preserve the reality of our own existence). A difficult dilemma indeed! What is required is an approach to world which undercuts the "cancer"—namely, the traditional subject-object dichotomy.

The reason this endeavor to rediscover man as being in the world is so important is that it strikes directly at one of the most acute problems of modern human beings—namely, that they have *lost their world,* lost their experience of community. Kierkegaard, Nietzsche, and the existentialists who followed them perdurably pointed out that the two chief sources of modern Western man's anxiety and despair were, first, his loss of sense of being and, second, his loss of his world. The existential analysts believe there is much evidence that these prophets were correct and that twentieth-century Western man not only experiences an alienation from the human world about him but also suffers an inner, harrowing conviction of being estranged (like, say, a paroled convict) in the natural world as well.

The writings of Frieda Fromm-Reichmann and Sullivan describe the state of the person who has lost his world. These authors, and others like them, illustrate how the problems of loneliness, isolation, and alienation are being increasingly dealt with in psychiatric literature. The assumption would seem likely that there is an increase not only in awareness of these problems among psychiatrists and psychologists but also in the presence of the conditions themselves. Broadly speaking, the symptoms of isolation and alienation reflect the state of a person whose relation to the world has become broken. Some psychotherapists have pointed out that more and more patients exhibit schizoid features and that the "typical" kind of psychic problem in our day is not hysteria, as it was in Freud's time, but the schizoid type—that is to say, problems of persons who are detached, unrelated, lacking in affect, tending toward depersonalization, and covering up their problems by means of intellectualization and technical formulations.[2]

There is also plenty of evidence that the sense of isolation, the alienation of oneself from the world, is suffered not only by people in pathological conditions but by countless "normal" persons as well in our day. Riesman presents a good deal of sociopsychological data in his study *The Lonely Crowd* to demonstrate that the isolated, lonely, alienated character type is characteristic not only of neurotic patients but of people as a whole in our society and that the trends in that direction have been increasing over the past couple of decades. He makes the significant point that these people have only a *technical* communication with their world; his "outer-directed" persons (the type characteristic of our day) relate to everything from its technical, external side. Their orientation, for example, was not "I liked the play," but "The play was *well done,*" "the article *well written,*" and so forth. Other portrayals of this condition of personal isolation and alienation in our society are given by Fromm in *Escape from Freedom,* particularly with respect to sociopolitical considerations; by Karl Marx, particularly in relation to the dehumanization arising out of the tendency in modern capitalism to value everything in the external, object-centered terms of money; and by Tillich from the spiritual viewpoint. Camus's *The Stranger* and Kafka's *The Castle,* finally, are surprisingly similar illustrations of our point: each gives a vivid and gripping picture of a man who is a stranger in his world, a stranger to other people whom he seeks or pretends to love; he moves about in a state of homelessness, vagueness, and haze as though he had no direct sense connection with his world but were in a foreign country where he does not know the language and has no hope of learning it but is always doomed to wander in quiet despair, incommunicado, homeless, and a stranger.

Nor is the problem of this loss of world simply one of lack of interpersonal relations or lack of communication with one's fellows. Its roots reach below the social levels to an alienation from the natural world as well. It is a particular experience of isolation which has been called "epistemological loneliness."[3] Underlying the economic, sociological, and psychological aspects of alienation can be found a profound common denominator—namely, the alienation which is the ultimate consequence of four centuries of the outworking of the separation of man as subject from the objective world. This alienation has expressed itself for several centuries in Western man's passion to gain power *over* nature, but now shows itself in an estrangement from nature and a vague, unarticulated, and half-suppressed sense of despair of gaining any real relationship with the natural world, including one's own body.

These sentences may sound strange in this century of such apparent confidence in science. But let us examine the matter more closely. Straus points out that Descartes, the father of modern thought, held that ego and consciousness were separated from the world and from other persons.[4] That is to say, consciousness is cut off and stands alone. Sensations do not tell us anything directly about the outside world; they only give us inferential data. Descartes is commonly the whipping boy in these days and made to shoulder the blame for the dichotomy between subject and object; but he was only reflecting the spirit of his age and the underground tendencies in modern culture, about which he saw and wrote with beautiful clarity. The Middle Ages, Straus goes on to say, is commonly thought of as otherworldly in contrast to the "present world" concerns of modern man. But actually the medieval Christian's soul was considered, while it did exist in the world, to be really related to the world. Men experienced the world about them as directly real (*vide* Giotto) and the body as immediate and real (*vide* Saint Francis). Since Descartes, however, the soul and nature have had nothing to do with each other. Nature belongs exclusively to the realm of *res extensa,* to be understood mathematically. We know the world only indirectly, by inference. This sets the problem we have been wrestling with ever since, the full implications of which did not emerge until the last century. Straus points out how the traditional textbooks on neurology and physiology have accepted this doctrine and have endeavored to demonstrate that what goes on neurologically has only a "sign" relation to the real world. Only "unconscious inferences lead to the assumption of the existence of an outside world."

Readers interested in this history of ideas will recall the important and imposing symbol of the same situation in Leibnitz's famous doctrine that all reality consists of *monads.* The monads had no doors or windows opening to each other, each being separated, isolated. "Each single unit is lonely in itself, without any direct communication. The horror of this idea was overcome by the harmonistic presupposition that in every monad the whole world is potentially present and that the development of each individual is in a natural harmony with the development of all the others. This is the most profound metaphysical situation in the early periods of bourgeois civilization. It fitted this situation because there was still a common world, in spite of the increasing social atomization" (Paul Tillich, *The Protestant Era,* p. 246). This doctrine of "pre-established harmony" is a carryover of the religious idea of providence. The relation between the person and the world was somehow "preordained." Descartes, in similar vein, held that God—whose existence he believed he had proved—guaranteed the relation between consciousness and the world. The socio-historical situation in the expanding phases of the modern period were such that the "faith" of Leibnitz and Descartes *worked*—that is, it reflected the fact that there was still a common world. But now that God is not only "dead," but a requiem has been sung over His grave, the stark isolation and alienation inherent in the relation between man and the world has become apparent. To put the matter less poetically, when the humanistic and Hebrew-Christian values disintegrated along with the cultural phenomena we have discussed above, the inherent implications of the situation emerged.

Thus it is by no means accidental that modern man feels estranged from nature, that each consciousness stands off by itself, alone. This has been "built in" to our education and to some extent even into our language. It means that the overcoming of this situation of isolation is not a simple task and requires something much more fundamental than merely the rearrangement of some of our present ideas.

Let us now inquire how the existential analysts undertake to rediscover man as a being interrelated with his world and to rediscover world as meaningful to man. They hold that the person and his world are a unitary, structural whole; the phrase "being in the world" expresses precisely that. The two poles, self and world, are always dialectically related. Self implies world and world self; there is neither without the other, and each is understandable only in terms of the other. It makes no sense, for example, to speak of man in his world (though we often do) as primarily a *spatial* relation. The phrase "match *in* a box" does imply a spatial relation, but to speak of a man *in* his home or in his office or in a hotel at the seashore implies something radically different.[5]

A person's world cannot be comprehended by describing the environment, no matter how complex we make our description. As we shall see below, environment is only one mode of world; and the common tendencies to talk of a person *in* an environment or to ask what "influence the environment has upon him" are vast oversimplifications. Even from a biological viewpoint, Von Uexküll holds, one is justified in assuming as many environments *(Umwelten)* as there are animals; "there is not one space and time only," he goes on to say, "but as many spaces and times as there are subjects."[6] How much more would it not be true that the human being has his own world? Granted that this confronts us with no easy problem: for we cannot describe world in purely objective terms, nor is world to be limited to our subjective, imaginative participation in the structure around us, although that too is part of being in the world.

World is the structure of meaningful relationships in which a person exists and in the design of which he participates. Thus world includes the past events which condition my existence and all the vast variety of deterministic influences which operate upon me. But it is these *as I relate to them,* am aware of them, carry them with me, molding, inevitably forming, building them in every minute of relating. For to be aware of one's world means at the same time to be designing it.

World is not limited to the past determining events but includes also all the possibilities which open up before any person and are not simply given in the historical situation. World is thus not to be identified with "culture." It includes culture but a good deal more, such as *Eigenwelt* (the own world which cannot be reduced merely to an introjection of the culture), as well as all the individual's future possibilities.[7] "One would get some idea," Schachtel writes, "of the unimaginable richness and depth of the world and its possible meanings for man, if he knew all languages and cultures, not merely intellectually but with his total personality. This would comprise the historically knowable world of man, but not the infinity of future possibilities."[8] It is the "openness of world" which chiefly distinguishes man's world from the closed worlds of animals and plants. This does not deny the finiteness of life; we are all limited by death and old age and are subject to infirmities of every sort. The point, rather, is that these possibilities are given within the context of the contingency of existence. In a dynamic sense, indeed, these future possibilities are the most significant aspect of any human being's world. For they are the potentialities with which he "builds or designs world"—a phrase the existential therapists are fond of using.

World is never something static, something merely given which the person then "accepts" or "adjusts to" or "fights." It is rather a dynamic pattern which, so long as I possess self-consciousness, I am in the process of forming and designing. Thus Binswanger speaks of world as "that toward which the existence has climbed and according to which it has designed itself,"[9] and goes on to emphasize that whereas a tree or an animal is tied to its "blueprint" in relation to the environment, "human existence not only contains numerous possibilities of modes of being, but is precisely rooted in this manifold potentiality of being."

The important and very fruitful use the existential analysts make of analyzing the patient's "world" is shown in Roland Kuhn's study of Rudolf, the butcher boy who shot a prostitute.[10] As Kuhn summarizes the facts in the case:

> On March 23, 1939, Rudolf R., a twenty-one-year-old, hard-working, inconspicuous butcher boy, having no police record, shot a prostitute with the intent to kill.
>
> He had left his job in the morning, donned his Sunday clothes, purchased a pistol and ammunition, and gone to Zurich on a one-way ticket. There he roamed the streets all day, stopping at several taverns but without drinking much. At 5 P.M. he met a prostitute in a bar, accompanied her to her room, had intercourse with her, and, after they had both dressed again, fired the shot. She was hit by the bullet but only slightly injured. Shortly after the criminal act, Rudolf surrendered to the police.

Noting that Rudolf was in mourning in this period following the death of his father, Kuhn goes to considerable lengths to understand the "world of the mourner." Rudolf had avoided and repressed the act of mourning for his mother who died when he was four. He had lived since that age in a "mourning" world—that is, generally depressed and seemingly without any experiences of joy. After the death of his mother he had slept in her bed for several years, seemed to be searching for her, or searching for the opportunity to experience the mourning and get it out of his system. Kuhn quotes Rilke, "Killing is one of the forms of our wandering mourning." The case is too complex as Kuhn describes it to summarize here. But at the conclusion of the study the reader is left with a completeness of understanding of Rudolf's "world of the mourner" and his attempted murder of the prostitute as an act of mourning for his mother. I do not believe that this clarity could be gained by any method other than such a description of "the patient in his world."

NINE

The Three Modes of World

THE EXISTENTIAL ANALYSTS, as we have discovered, distinguish three modes of world—that is, three simultaneous aspects of world—which characterize the existence of each one of us as being in the world. First, there is *Umwelt*, literally meaning "world around"; this is the biological world, generally called in our day the environment. There is, second, the *Mitwelt*, literally the "with world," the world of beings of one's own kind, the world of one's fellow men. The third is *Eigenwelt*, the "own world," the world of relationship to oneself.

The first, *Umwelt*, is what is taken in general parlance as world—namely, the world of objects about us, the natural world. All organisms have an *Umwelt*. For animals and human beings the *Umwelt* includes biological needs, drives, instincts —the world one would still exist in if, let us hypothesize, one had no self-consciousness. It is the world of natural law and natural cycles, of sleep and awakeness, of being born and dying, desire and relief, the world of finiteness and biological determinism, the "thrown world" into which each of us was hurled by our birth and to which each of us must in some way adjust. The existential analysts do not at all neglect the reality of the natural world; "natural law is as valid as ever," as Kierkegaard put it. They have no truck with the idealists who would reduce the material world to an epiphenomenon or with the intuitionists who would make it purely subjective or with anyone who would underestimate the importance of the world of biological determinism. Indeed, their insistence on taking the objective world of nature seriously is one of their distinctive characteristics. In reading them I often have the impression that they are able to grasp the *Umwelt,* the material world, with greater reality than those who segment it into "drives" and "substances," precisely because they are not limited to *Umwelt* alone, but see it also in the context of human self-consciousness.[1] Boss's understanding of the patient with the "feces and church steeple" dream cited above is an excellent example. They insist strongly that it is an oversimplification and radical error to deal with human beings as though *Umwelt* were the only mode of existence or to carry over the categories which fit *Umwelt* to make a Procrustean bed upon which to force all human experience. In this connection, the existential analysts are *more empirical*—that is, more respectful of actual human phenomena—than the mechanists or positivists or behaviorists.

The *Mitwelt* is the world of interrelationships with human beings. But it is not to be confused with "the influence of the group upon the individual," or "the collective mind," or the various forms of "social determinism." The distinctive quality of *Mitwelt* can be seen when we note the difference between a herd of animals and a community of people. Howard Liddell has pointed out that for his sheep the "herd instinct consists of keeping the environment constant." Except in mating and suckling periods, a flock of collie dogs or children will do as well for the sheep providing such an environment is kept constant. In a group of human beings, however, a vastly more complex interaction goes on, with the meaning of the others in the group partly determined by one's own relationship to them.

Strictly speaking, we should say animals have an *environment,* human beings have a *world.* For world includes the structure of meaning which is designed by the interrelationship of the persons in it. Thus the meaning of the group for me depends in part upon how I put myself into it. And thus, also, love can never be understood on a purely biological level but depends upon such factors as personal decision and commitment to the other person.[2]

The categories of "adjustment" and "adaptation" are entirely accurate in *Umwelt.* I adapt to the cold weather and I adjust to the periodic needs of my body for sleep; the critical point is that the weather is not changed by my adjusting to it nor is it affected at all. Adjustment occurs between two objects, or a person and an object. But in *Mitwelt,* the categories of adjustment and adaptation are not accurate; the term *relationship* offers the right category. If I insist that another person adjust to me, I am not taking him as a person, as *Dasein,* but as an instrumentality; and even if I adjust to myself, I am using myself as an object. One can never accurately speak of human beings as "sexual objects"; once a person is a sexual object, you are not talking about a person any more. *The essence of relationship is that in the encounter both persons are changed.* Providing the human beings involved are not too severely ill and have some degree of consciousness, relationship always involves mutual awareness; and this already is the process of being mutually affected by the encounter.

The *Eigenwelt,* or "own world," is the mode which is least adequately dealt with or understood in modern psychology and depth psychology; indeed, it is fair to say that it is almost ignored. *Eigenwelt* presupposes self-awareness, self-relatedness, and is uniquely present in human beings. But it is not merely a subjective, inner experience; it is rather the basis on which we see the real world in its true perspective, the basis on which we relate. It is a grasping of what something in the world—this bouquet of flowers, this other person—means to *me.* Suzuki has remarked that in Eastern languages, such as Japanese, adjectives always include the implication of "for-me-ness." That is to say, "this flower is beautiful" means "*for me* this flower is beautiful." Our Western dichotomy between subject and object has led us, in contrast, to assume that we have said most if we state that the flower is beautiful entirely divorced from ourselves, as though a statement were the more true in proportion to how little we ourselves have to do with it! This leaving of *Eigenwelt* out of the picture not only contributes to arid intellectualism and loss of vitality but obviously also has much to do with the fact that modern people tend to lose the sense of reality of their experiences.

It should be clear that these three modes of world are always interrelated and always condition each other. At every moment, for example, I exist in *Umwelt,* the biological world. But how I relate to my need for sleep or the weather or any instinct —how, that is, I see in my own self-awareness this or that aspect of *Umwelt*—is crucial for its meaning for me and conditions how I will react to it. The human being lives in *Umwelt, Mitwelt,* and *Eigenwelt* simultaneously. They are by no means three different worlds but three simultaneous modes of being in the world.

Several implications follow from the above description of the three modes of world. One is that the reality of being in the world is lost if *one of these modes is emphasized to the exclusion of the other two.* In this connection, Binswanger holds that classical psychoanalysis deals only with the *Umwelt.* The genius and the value of Freud's work lies in uncovering man in the *Umwelt,* the mode of instincts, drives, contingency, biological determinism. But traditional psychoanalysis has only a shadowy concept of *Mitwelt,* the mode of the interrelation of persons as subjects. One might argue that such psychoanalysis does have a *Mitwelt* in the sense that individuals need to find each other for the sheer necessity of meeting biological needs, that libidinal drives require social outlets and make social relationships necessary. But this is simply to derive *Mitwelt* from *Umwelt,* to make *Mitwelt* an epiphenomenon of *Umwelt;* and it means that we are not really dealing with *Mitwelt* at all but only another form of *Umwelt.*

It is clear that the interpersonal schools do have a theoretical basis for dealing directly with *Mitwelt.* This is shown, to take only one example, in Sullivan's interpersonal theory. Though they should not be identified, *Mitwelt* and interpersonal theory have a great deal in common. The danger at this point, however, is that if *Eigenwelt* in turn is omitted, interpersonal relations tend to become hollow and sterile. It is well known that Sullivan argued against the concept of the individual personality, and made a great effort to define the self in terms of "reflected appraisal" and social categories—i.e., the roles the person plays in the interpersonal world.[3] Theoretically, this suffers from considerable logical inconsistency and indeed goes directly against other very important contributions of Sullivan. Practically, it tends to make the self a mirror of the group around one, to empty the self of vitality and originality, and to reduce the interpersonal world to mere "social relations." It opens the way to the tendency which is directly opposed to the goals of Sullivan and other interpersonal thinkers—namely, social conformity. *Mitwelt* does not automatically absorb either *Umwelt* or *Eigenwelt.*

But when we turn to the mode of *Eigenwelt* itself, we find ourselves on the unexplored frontier of psychotherapeutic theory. What does it mean to say "the self in relation to itself"? What goes on in the phenomena of consciousness, of self-awareness? What happens in "insight" when the inner Gestalt of a person reforms itself? Indeed, what does the "self knowing itself" mean? Each of these phenomena goes on almost every instant with all of us; they are indeed closer to us than our breathing. Yet, perhaps precisely because they are so near to us, no one knows what is happening in these events. This mode of the self in relation to itself was the aspect of experience which Freud never really saw, and it is doubtful whether any school has as yet achieved a basis for adequately dealing with it. *Eigenwelt* is certainly the hardest mode to grasp in the face of our Western technological preoccupations. It may well be that the mode of *Eigenwelt* will be the area in which most clarification will occur in the next decades.

Another implication of this analysis of the modes of being in the world is that it gives us a basis for the psychological understanding of love. The human experience of love obviously cannot be adequately described within the confines of *Umwelt*. The interpersonal schools, at home chiefly in *Mitwelt*, have dealt with love, particularly in Sullivan's concept of the meaning of the "chum" and in Fromm's analysis of the difficulties of love in contemporary estranged society. But there is reason for doubting whether a theoretical foundation for going further is yet present in these or other schools. The same general caution given above is pertinent here—namely, that without an adequate concept of *Umwelt*, love becomes empty of vitality, and without *Eigenwelt*, it lacks power and the capacity to fructify itself.

In any case, *Eigenwelt* cannot be omitted in the understanding of love. Nietzsche and Kierkegaard continually insisted that to love presupposes that one has already become the "true individual," the "Solitary One," the one who "has comprehended the deep secret that also in loving another person one must be sufficient unto oneself."[4] They, like other existentialists, do not attain to love themselves; but they help perform the psycho-surgical operations on nineteenth-century man which may clear blockages away and make love possible. By the same token, Binswanger and other existential therapists speak frequently of love. And though one could raise questions about how love is actually dealt with by them in given therapeutic cases, they nonetheless give us the theoretical groundwork for ultimately dealing with love adequately in psychotherapy.

One feels in many of the psychological and psychiatric discussions of love in America a lack of the *tragic* dimension. Indeed, to take tragedy into the picture in any sense requires that the individual be understood in the three modes of world —the world of biological drive, fate, and determinism *(Umwelt);* the world of responsibility to fellow men *(Mitwelt);* and the world in which the individual can be aware *(Eigenwelt)* of the fate he alone at that moment is struggling with. The *Eigenwelt* is essential to any experience of tragedy, for the individual must be conscious of his own identity in the midst of the vast natural and social forces operating a destiny upon him. It has been rightly said that we lack a sense of tragedy in America—and hence produce few real tragedies in drama or other forms of art—because we lack the sense of the individual's own identity and consciousness.

TEN

Of Time and History

THE NEXT CONTRIBUTION of the existential analysts we shall consider is their distinctive approach to *time.* They are struck by the fact that the most profound human experiences, such as anxiety, depression, and joy, occur more in the dimension of time than in space. They boldly place time in the center of the psychological picture and proceed to study it not in the traditional way as an analogy to space but in its own existential meaning for the patient.

An example of the fresh light this new approach to time throws upon psychological problems is seen in an engaging case study by Eugene Minkowski.[1] Coming to Paris after his psychiatric training, Minkowski was struck by the relevance of the time dimension, then being proclaimed by Bergson, to the understanding of psychiatric patients.[2] Minkowski lived with this patient. In his own words:

> In the year 1922, a stroke of good luck—or, more exactly, life's vicissitudes—obliged me to spend two months as the personal physician of a patient. . . .
>
> The patient, a man of sixty-six who presented a depressive psycho-

sis and delusions of persecution, expressed thoughts of guilt and ruin. . . . People looked oddly at him in the street, his servants were paid to spy on him and betray him, every newspaper article was directed at him, and books had been printed solely against him and his family. . . .

I had the possibility of following him from day to day, not in a mental hospital or sanitarium, but in an ordinary environment. . . . I was not only able to observe the patient but also at almost each instant I had the possibility of comparing his psychic life and mine. It was like two melodies being played simultaneously; although these two melodies are as unharmonious as possible, nevertheless, a certain balance becomes established between the notes of the one and the other and permits us to penetrate a bit more deeply into our patient's psyche.

. . . *Where, exactly, is the discordance between his psyche and our own?* We ask, what is a delusion? Is it really nothing but a disorder of perception and of judgment? This brings us back to our present problem—namely, where is the discordance between the patient's psyche and our own?

When I arrived, he stated that his execution would certainly take place that night; in his terror, unable to sleep, he also kept me awake all that night. I comforted myself with the thought that, come the morning, he would see that all his fears had been in vain. However, the same scene was repeated the next day and the next, until after three or four days I had given up hope, whereas his attitude had not budged one iota. What had happened? It was simply that I, as a normal human being, had rapidly drawn from the observed facts my conclusions about the future. I now knew that he would continue to go on, day after day, swearing that he was to be tortured to death that night, and so he did, giving no thought to the present or the past. . . . This carry-over from past and present into the future was completely lacking in him.

. . . This reasoning . . . indicated a profound disorder in his general attitude toward the future; that time which we normally integrate into a progressive whole was here split into isolated fragments.

In this case, Minkowski points out that the patient *could not relate to time* and that each day was a separate island with no past and no future. Traditionally, the therapist would reason simply that the patient cannot relate to the future, cannot "temporize," *because* he has these delusions. Minkowski proposes the exact opposite. "Could we not," he asks, "on the contrary suppose *the more basic disorder is the distorted attitude toward the future, while the delusion is only one of its manifestations?*" Minkowski's original approach throws a beam of illumination on these dark, unexplored areas of time and introduces a new freedom from the limits and shackles of clinical thought when bound only to traditional ways of thinking.

As I was reading this case, a parallel to my own practice in psychotherapy came to mind. That is, I discovered with some surprise that if we can help the severely anxious or depressed patient to focus on some point in the future when he will be *outside* his anxiety or depression, the battle is half won. The essence of severe anxiety and depression is that it engulfs our whole selves, it feels as Minkowski says, *universal.* But the class that the patient is teaching which makes him so anxious will be over, or the dreaded session with his boss will be passed—what will he feel then? This focusing upon some point in time *outside* the depression or anxiety gives the patient a perspective, a view from on high so to speak; and this may well break the chains of the anxiety or depression. The patient may then relax, and some hope creeps in.

If we psychotherapists, in our preoccupation with the content of depression or anxiety, have forgotten the time dimension, certainly the poets, often nearer to the actual existential experiences than we, have not. In the famous lines, Shakespeare has his acutely depressed character Macbeth ponder not his crime, which would be the *content* dimension, but the *time* dimension:

> Tomorrow, and tomorrow, and tomorrow,
> Creeps in the petty pace from day to day,
> To the last syllable of recorded time;

And all our yesterdays have lighted fools
The way to dusty death.

This new approach to time begins with observing that the most crucial fact about existence is that it *emerges*—that is, it is always in the process of becoming, always developing in time, and is never to be defined as static points.[3] The existential therapists propose a psychology literally of *being,* rather than "is" or "has been" or fixed inorganic categories. Though their concepts were worked out several decades ago, it is highly significant that experimental work in psychology, such as that by Mowrer and Liddell, illustrates and bears out their conclusions. At the end of one of his most important papers, Mowrer holds that time is the distinctive dimension of human personality. "Time binding"—that is, the capacity to bring the past into the present as part of the total causal nexus in which living organisms act and react, together with the capacity to act in the light of the long-term future—is "the essence of mind and personality alike."[4] Liddell has shown that his sheep can keep time—anticipate punishment—for about fifteen minutes and his dogs for about half an hour; but a human being can bring the past of thousands of years ago into the present as data to guide his present actions. And he can likewise project himself in conscious imagination into the future not only for a quarter of an hour but for weeks and years and decades. This capacity to transcend the immediate boundaries of time, to see one's experience self-consciously in the light of the distant past and the future, to act and react in these dimensions, to learn from the past of a thousand years ago and to mold the long-time future, is the unique characteristic of human existence.

The existential therapists agree with Bergson that "time is the heart of existence" and that our error has been to think of ourselves primarily in spatialized terms appropriate to *res extensa,* as though we were objects which could be located like substances at this spot or that. By this distortion we lose our genuine and real existential relation with ourselves, and indeed with other persons around us. As a consequence of this overemphasis on spatialized thinking, says Bergson, "the moments when we grasp ourselves are rare, and consequently we are seldom free."[5] Or, when we have taken time into the picture, it has been in the sense of Aristotle's definition, the dominant one in the tradition of Western thought, "For the time is this: what is counted in the movement in accordance with what is earlier and later."

The striking thing about this description of "clock time" is that it really is an analogy from space, and one can best understand it by thinking in terms of a line of blocks or regularly spaced points on a clock or calendar. This approach to time is most fitting in the *Umwelt,* where we view the human being as an entity set among the various conditioning and determining forces of the natural world and acted upon by instinctual drives. But in the *Mitwelt,* the mode of personal relations and love, quantitative time has much less to do with the significance of an occurrence; the nature or degree of one's love, for example, can never be measured by the number of years one has known the loved one. It is true that clock time has much to do with *Mitwelt:* many people sell their time on an hourly basis, and daily life runs on schedules. We refer rather to the inner meaning of the events. "No clock strikes for the happy one," says a German proverb quoted by Straus. Indeed, the most signifiant events in a person's psychological existence are likely to be precisely the ones which are "immediate," breaking through the usual steady progression of time.

Finally, the *Eigenwelt,* the own world of self-relatedness, self-awareness, and insight into the meaning of an event for one's self, has practically nothing whatever to do with Aristotle's clock time. The essence of self-awareness and insight are that they are "there"—instantaneous, immediate—and the moment of awareness has its significance for all time. One can see this easily by noting what happens in oneself at the instant of an insight or any experience of grasping oneself; the insight occurs with suddenness, is "born whole," so to speak. And one will discover that, though meditating on the insight for an hour or so may reveal many of its further implications, the insight is not clearer—and disconcertingly enough, often not as clear —at the end of the hour as it was at the beginning.

The existential therapists also observed that the most profound psychological experiences are peculiarly those which shake the individual's relation to time. Severe anxiety and depression blot out time, annihilate the future. Or, as Minkowski proposes, it may be that the disturbance of the patient in relation to time, his inability to "have" a future, gives rise to his anxiety and depression. In either case, the most painful aspect of the sufferer's predicament is that he is unable to imagine a future moment in time when he will be out of the anxiety or depression. We see a similar close interrelationship between the disturbance of the time function and neurotic symptoms. Repression and other processes of the blocking off of awareness are in essence methods of ensuring that the usual relation of past to present will not obtain. Since it would be too painful or in other ways too threatening for the individual to retain certain aspects of his past in his present consciousness, he must carry the past along like a foreign body *in* him but not *of* him, as it were, an encapsulated fifth column which thereupon compulsively drives to its outlets in neurotic symptoms.

However one looks at it, the problem of time has a peculiar importance in understanding human existence. The reader may agree at this point but feel that, if we try to understand time in other than spatial categories, we are confronted with a mystery. He may well share the perplexity of Augustine, who wrote, "When no one asks me what time is, I know, but when I would give an explanation of it in answer to a man's question I do not know."[6]

One of the distinctive contributions of the existential analysts to this problem is that, having placed time in the center of the psychological picture, they then propose that the *future*, in contrast to present or past, is the dominant mode of time for human beings. Personality can be understood only as we see it on a trajectory toward its future; a man can understand himself only as he projects himself forward. This is a corollary of the fact that the person is always becoming, always emerging into the future. The self is to be seen in its potentiality. *"A self, every instant it exists," Kierkegaard wrote, "is in process of becoming, for the self . . . is only that which it is to become."* The existentialists do not mean "distant future" or anything connected with using the future as an escape from the past or present; they mean only to indicate that the human being, so long as he possesses self-awareness and is not incapacitated by anxiety or neurotic rigidities, is always in a dynamic self-actualizing process, always exploring, molding himself, and moving into the immediate future.

They do not neglect the past, but they hold it can be understood only in the light of the future. The past is the domain of *Umwelt,* of the contingent, natural historical, deterministic forces operating upon us. But since we do not live exclusively in *Umwelt,* we are never merely the victims of automatic pressures from the past. *The deterministic events of the past take their significance from the present and future.* As Freud put it, we are anxious *lest* something happen in the future. "The word of the past is an oracle uttered," remarked Nietzsche. "Only as builders of the future, as knowing the present, will you understand it." All experience has a historical character, but the error is to treat the past in mechanical terms. The past is not the "now which was," nor any collection of isolated events, nor a static reservoir of memories or past influences or impressions. The past, rather, is the domain of contingency in which we accept events and from which we select events in order to fulfill our potentialities and to gain satisfactions and security in the immediate future. This realm of the past, of natural history and "thrownness," Binswanger points out, is the mode which classical psychoanalysis has, *par excellence,* made its own for exploration and study.

But as soon as we consider the exploration of a patient's past in psychoanalysis, we note two very curious facts. First is the obvious phenomenon observed every day, that the events in the past which the patient remembers have very little, if any, necessary connection with the quantitative events that actually happened to him as a child. One single thing that occurred to him at a given age is remembered and thousands of things are forgotten, and even the events that occurred most frequently, like getting up in the morning, are most apt to leave no impression. Alfred Adler pointed out that memory is a creative process, that we remember what has significance for our "style of life," and that the whole "form" of memory is, therefore, a mirror of the individual's style of life. What an individual seeks *to become* determines what he remembers of his *has been.* In this sense the future determines the past.

The second fact is this: *whether or not a patient can even recall the significant events of the past depends upon his decision with regard to the future.* Every therapist knows that patients may bring up past memories *ad interminum* without any memory ever moving them, the whole recital being flat, inconsequential, tedious. From an existential point of view, the problem is not at all that these patients happened to have endured impoverished pasts; it is rather that they cannot or do not commit themselves to the present and future. Their past does not become alive because nothing matters enough to them in the future. Some hope and commitment to work toward changing something in the immediate future, be it overcoming anxiety or other painful symptoms or integrating oneself for further creativity, is necessary before any uncovering of the past will have reality.

One practical implication of the above analysis of time is that psychotherapy cannot rest on the usual automatic doctrines of historical progress. The existential analysts take history very seriously,[7] but they protest against any tendency to evade the immediate, anxiety-creating issues in the present by taking refuge behind the determinism of the past. They are against the doctrines that historical forces carry the individual along automatically, whether these doctrines take the form of the religious beliefs of predestination or providence, the deteriorated Marxist doctrine of historical materialism, the various psychological doctrines of determinism, or that most common form of such historical determinism in our society, faith in automatic technical progress. Kierkegaard was very emphatic on this point:

> Whatever the one generation may learn from the other, that which is genuinely human no generation learns from the foregoing. . . . Thus no generation has learned from another to love, no generation begins at any other point than at the beginning, no generation has a shorter task assigned to it than had the previous generation. . . . In this respect every generation begins primitively, has no different task from that of every previous generation, nor does it get further, except in so far as the preceding generation shirked its task and deluded itself.[8]

This implication is particularly relevant to psychotherapy, since the popular mind so often makes of psychoanalysis and other forms of psychotherapy the new technical authority which will take over for them the burden of learning to love. Obviously all any therapy can do is to help a person remove the blocks which keep him from loving. It cannot love for him, and it is doing him ultimate harm if it dulls his own responsible awareness at this point.

A last contribution of this existential analysis of time lies in its understanding of the process of insight. Kierkegaard uses the engaging term *Augenblick,* literally meaning the "blinking of an eye" and generally translated "the pregnant moment." It is the moment when a person suddenly grasps the meaning of some important event in the past or future in the present. Its pregnancy consists of the fact that it is never an intellectual act alone. The grasping of the new meaning always presents the possibility and necessity of some personal decision, some shift in Gestalt, some new orientation of the person toward the world and future. This is experienced by most people as the moment of most heightened awareness; it is referred to in psychological literature as the "aha" experience. On the philosophical level, Paul Tillich describes it as the moment when "eternity touches time," for which moment he has developed the concept of *Kairos,* "time fulfilled." In religion and literature this moment when eternity touches time is known as an epiphany.

ELEVEN

Transcending the Immediate Situation

A FINAL CHARACTERISTIC of man's existence *(Dasein)* which we shall discuss is the capacity to transcend the immediate situation. If one tries to study the human being as a composite of substances, one does not need to deal with the disturbing fact that existence is always in process of self-transcending. But if we are to understand a given person as existing, dynamic, at every moment becoming, we cannot avoid this dimension. This capacity is already stated in the term *exist*—that is, "to stand out from." Existing involves a continual emerging, in the sense of emergent evolution, a transcending of one's past and present in terms of the future. Thus *transcendere*—literally "to climb over or beyond"—describes what every human being is engaged in doing every moment when he is not seriously ill or temporarily blocked by despair or anxiety. One can see this emergent evolution in all life processes. Nietzsche has Zarathustra proclaim, "And this secret spake Life herself to me, 'Behold' said she, 'I am that which must ever surpass itself.' " But it is much more radically true of human existence, where the capacity for self-awareness qualitatively increases the range of consciousness and, therefore, greatly enlarges the range of possibilities of transcending the immediate situation.

The term *transcending* is open to much misunderstanding and indeed often calls forth violent antagonism.[1] In this country the term is relegated to vague and ethereal things which, as Bacon remarked, are better dealt with in "poesy, where transcendences are more allowed," or associated with Kantian a priori assumptions, or with New England transcendentalism or religious other worldliness, or with things unempirical and unrelated to actual experience. We mean something different from all of these.

It has been suggested, furthermore, that the word has lost its usefulness and another should be found. That would be fine if another were available that would adequately describe the exceedingly important immediate human experience to which this term, when used by Goldstein and the existential writers, refers; for any adequate description of human beings requires that the experience be taken into account. Some suspicion of the term obviously is sound to the extent that the word can serve to elevate any given topic out of any immediate field in which it can be discussed and thus lead to obscurantism. And often in such cases the transcendent thing aimed at is given a capital letter, such as Self or Wholeness, apparently to bootleg in some quality of divinity. It must be confessed that occasional usages of the term in existential literature have a similar effect, particularly when the "transcendental categories" of Husserl are assumed without explanation of how they apply. Other objections to the term, less justifiable, may arise from the fact that the capacity to transcend the present situation introduces a disturbing fourth dimension, a *time* dimension, and this is a serious threat to the traditional way of describing human beings in terms of static substances. The term is likewise rejected by those who seek to make no distinction between animal and human behavior or to understand human psychology in terms only of mechanical models. This capacity we are about to discuss does in actual fact present difficulties to those approaches since it is uniquely characteristic of human beings.

The neurobiological base for this capacity is classically described by Kurt Goldstein. Goldstein found that his brain-injured patients—chiefly soldiers with portions of the frontal cortex shot away—had specifically lost the ability to abstract, to think in terms of "the possible." They were tied to the immediate concrete situation in which they found themselves. When their closets happened to be in disarray, they were thrown into profound anxiety and disordered behavior. They exhibited compulsive orderliness—which is a way of holding oneself at every moment rigidly to the concrete situation. When asked to write their names on a sheet of paper, they would typically write in the very corner, any venture out from the specific boundaries of the edges of the paper representing too great a threat. It was as though they were threatened with dissolution of the self unless they remained related at every moment to the immediate situation, as though they could "be a self" only as the self was bound to the concrete items in space. Goldstein holds that the distinctive capacity of the normal human being is precisely this capacity to abstract, to use symbols, to orient oneself beyond the immediate limits of the given time and space, to think in terms of "the possible." The injured, or "ill," patients were characterized by loss of range of possibility. Their world space was shrunk, their time curtailed, and they suffered a consequent radical loss of freedom.

The capacity of the normal human being to transcend the present situation is exemplified in all kinds of behavior. One is the capacity to transcend the boundaries of the present moment in time—as we pointed out in our discussion above—and to bring the distant past and the long-term future into one's immediate existence. It is also exemplified in the human being's unique capacity to think and talk in symbols. Reason and the use of symbols are rooted in the capacity to stand outside the particular object or sound at hand, say these boards on which my typewriter sits and the two syllables that make up the word "table," and agreeing with each other that these will stand for a whole class of objects.

The capacity is particularly shown in social relationships, in the normal person's relation to the community. Indeed, the whole fabric of trust and responsibility in human relations presupposes the capacity of the individual to "see himself as others see him," as Robert Burns puts it in contrasting himself with the field mouse, to see himself as the one fulfilling his fellow men's expectations, acting for their welfare or failing to. Just as this capacity for transcending the situation is impaired with respect to the *Umwelt* in the brain-injured, as we saw in Goldstein's patients, it is impaired with respect to the *Mitwelt* in the psychopathic disorders. These are the disorders of those in whom the capacity to see themselves as others see them is absent or does not carry sufficient weight, who are then said to lack "conscience." The term *conscience,* significantly enough, is in many languages the same word as *consciousness,* both meaning "to know with." Nietzsche remarked, "Man is the animal who can make promises." By this he did not mean promises in the sense of social pressure or simply introjection of social requirement (which are oversimplified ways of describing conscience, errors which arise from conceiving of *Mitwelt* apart from *Eigenwelt).* Rather, he meant that an individual can be aware of the fact that he has given his word, can see himself as the one who makes the agreement. Thus, to make promises presupposes conscious self-relatedness and is a very different thing from simple conditioned "social behavior," acting in terms of the requirements of the group or herd or hive. In the same light, Sartre writes that dishonesty is a uniquely human form of behavior: "The lie is a behavior of transcendence."

It is significant at this point to note the great number of terms used in describing human actions which contain the prefix *re—re*-sponsible, *re*-collect, *re*-late, and so on. In the last analysis, all imply and rest upon this capacity to "come back" to oneself as the one performing the act. This is illustrated with special clarity in the peculiarly human capacity to be *responsible* (a word combining *re* and *spondere,* "promise"), designating the one who can be depended upon, who can promise to give back, to answer. Erwin Straus describes man as "the questioning being," the organism who at the same moment that he exists can question himself and his own existence.[2] Indeed, the whole existential approach is rooted in the always curious phenomenon that we have in man a being who not only *can* but *must,* if he is to realize himself, question his own being. One can see at this point that the discussion of dynamisms of social adjustment, such as "introjection," "identification," and so forth, is oversimplified and inadequate when it omits the central fact of all—namely, the person's capacity to be aware at the moment that he is the one responding to the social expectation, the one choosing (or not choosing) to guide himself according to a certain model. This is the distinction between rote social conformity on one hand and the freedom, originality, and creativity of genuine social response on the other. The latter are the unique mark of the human being acting in the light of "the possible."

Self-consciousness implies self transcendence. The one has no reality without the other. It will have become apparent to many readers that the capacity to transcend the immediate situation uniquely presupposes *Eigenwelt*—that is, the mode of behavior in which a person sees himself as subject and object at once. The capacity to transcend the situation is an inseparable part of self-awareness, for it is obvious that the mere awareness of oneself as a being in the world implies the capacity to stand outside and look at oneself and the situation and to assess and guide oneself by an infinite variety of possibilities. The existential analysts insist that the human being's capacity for transcending the immediate situation is discernible in the very center of human experience and cannot be sidestepped or overlooked without distorting and making unreal and vague one's picture of the person. This is particularly cogent and true with respect to data we encounter in psychotherapy. All of the peculiarly neurotic phenomena, such as the split of unconsciousness from consciousness, repression, blocking of awareness, self-deceit by way of symptoms, *ad interminum,* are misused, "neurotic" forms of the fundamental capacity of the human being to relate to himself and his world as subject and object at the same time. As Lawrence Kubie has written, "The neurotic process is always a symbolic process: and the split into parallel yet interacting streams of conscious and unconscious processes starts approximately as the child begins to develop the rudiments of speech. . . . It may be accurate to say, therefore, that the neurotic process is the price that we pay for our most precious human heritage, namely our ability to represent experience and communicate our thoughts by means of symbols."[3] The essence of the use of symbols, we have tried to show, is the capacity to transcend the immediate, concrete situation.

We can now see why Medard Boss and the other existential psychiatrists and psychologists make this capacity to transcend the immediate situation the basic and unique characteristic of human existence. "Transcendence and being-in-the-world are names for the identical structure of Dasein, which is the foundation for every kind of attitude and behavior."[4] Boss goes on in this connection to criticize Binswanger for speaking of different kinds of "transcendences"—the "transcendence of love" as well as the "transcendence of care." This unnecessarily complicates the point, says Boss; and it makes no sense to speak of "transcendences" in the plural. We can only say, holds Boss, that man has the capacity for transcending the immediate situation because he has the capacity for *Sorge*—that is, for "care" or, more accurately, for understanding his being and taking responsibility for it. (This term *Sorge* is from Heidegger and is basic to existential thought; it is used often in the form of *Fürsorge,* meaning "care for," "concerned for the welfare of.") *Sorge* is for Boss the encompassing notion and includes love, hate, hope, and even indifference. All attitudes are ways of behaving in *Sorge* or lack of it. In Boss's sense the capacity of man to have *Sorge* and to transcend the immediate situation are two aspects of the same thing.

We need now to emphasize that this capacity to transcend the immediate situation is not a "faculty" to be listed along with other faculties. It is rather given in the ontological nature of being human. To abstract, to objectivate, are evidences of it. But, as Heidegger puts it, "transcendence does not consist of objectivation, but objectivation presupposes transcendence." That is to say, the fact that the human being can be self-related gives him, as one manifestation, the capacity to objectify his world, to think and talk in symbols, and so forth. This is Kierkegaard's point when he reminds us that to understand the self we must see clearly that "imagination is not one faculty on a par with others, but, if one would so speak, it is the faculty *instar omnium* [for all faculties]. What feeling, knowledge or will a man has depends in the last resort upon what imagination he has, that is to say, upon how these things are reflected. . . . Imagination is the possibility of all reflection, and the intensity of this medium is the possibility of the intensity of the self."[5]

It remains to make more specific what is implicit above—namely, that this capacity for transcending the immediate situation is the basis of human freedom. The unique characteristic of the human being is the vast range of possibilities in any situation, which in turn depends upon his self-awareness, his capacity to run through in imagination the different ways of reacting in a given situation. Von Uexküll, in his metaphor of the forest, contrasts the different environments that creatures with different purposes find in the same forest. The insect in the tree has one environment; the romantic girl who walks in the forest has another; the woodsman who comes to chop down trees for firewood has another; the artist who comes to paint the forest has still another.[6] Binswanger uses this metaphor to illustrate the variety of ways the human being can select among many self-world relationships. All of these depend upon our imagination-based possibilities of transcending—i.e., putting our own purposes upon the scene before us. In a variety of ways the human being can select among many self-world relationships. The "self" is the capacity to see oneself in these many possibilities.

This freedom with respect to world, Binswanger goes on to point out, is the mark of the psychologically healthy person; to be rigidly confined to a specific "world," as was Ellen West, is the mark of psychological disorder. What is essential is "freedom in designing world" or "letting world occur," as Binswanger puts it. "So deeply founded," he observes, "is the essence of freedom as a necessity in existence that it can also dispense with existence itself."[7]

TWELVE

Concerning Therapeutic Technique

MOST OF THOSE who read works on existential analysis as handbooks of technique are bound to be disappointed. They will not find specifically developed practical methods.[1] Many of the existential analysts are not greatly concerned with technical matters. The basic reason for the fact that these psychiatrists and psychologists are not so concerned with formulating technique, and make no apologies for this fact, is that existential analysis is a way of understanding human existence rather than a system of "how tos." Its representatives believe that one of the chief (if not *the* chief) blocks to the understanding of human beings in Western culture is precisely the overemphasis on technique, an overemphasis that goes along with the tendency to see the human being as an object to be calculated, managed, "analyzed."[2] Our Western tendency has been to believe that *understanding follows technique;* if we get the right technique, then we can penetrate the riddle of the patient, or, as said popularly with amazing perspicacity, we can "get the other person's number." The existential approach holds the exact opposite—namely, that *technique follows understanding.* The central task and responsibility of the therapist is to seek to understand the patient as a being and as being in his world. All technical problems are subordinate to this understanding. Without this understanding, technical facility is at best irrelevant, at worst a method of "structuralizing" the neurosis. With it, the groundwork is laid for the therapist's being able to help the patient recognize and experience his own existence, and this is the central process of therapy. This does not derogate disciplined technique. It rather puts it into perspective.

It is clear at the outset that what distinguishes existential therapy is not what the therapist would specifically do, let us say, in meeting anxiety or confronting resistance or getting the life history and so forth, but rather the *context* of his therapy. How an existential therapist might interpret a given dream or an outburst of temper on the patient's part might not differ from what a classical psychoanalyst might say, if each were taken in isolated fashion. But the context of existential therapy would be very distinct. It would always focus on the questions of how this dream throws light on this particular patient's existence in his world, what it says about *where* he is at the moment and what he is moving toward, and so forth. The context is the patient not as a set of psychic dynamisms or mechanisms but as a human being who is choosing, committing, and pointing himself toward something right now. The context is dynamic, immediately real, and present.

I shall try to block out some implications concerning therapeutic technique from my knowledge of the works of the existential therapists and from my own experience of how their emphases have contributed to me, a therapist trained in psychoanalysis in its broad sense. Making a systematic summary would be presumptuous to try and impossible to accomplish, but I hope the following points will at least suggest some of the important therapeutic implications. It should be clear at every point, however, that the really important contributions of this approach are its deepened understanding of human existence, and one gets no place talking about isolated techniques of therapy unless the understanding we have sought to give in the earlier portions of these chapters is presupposed at every point.

The first implication is the variability of techniques among the existential therapists. Boss, for example, uses couch and free association in traditional Freudian manner and permits a good deal of acting out of transference. Others would vary as much as the different schools vary anyway. But the critical point is that the existential therapists have a definite reason for using any given technique with a given patient. They sharply question the use of techniques simply because of rote, custom, or tradition. Their approach also does not at all settle for the air of vagueness and unreality that surrounds many therapeutic sessions, particularly in the eclectic schools which allegedly have freed themselves from bondage to a traditional technique and select from all schools as though the presuppositions of these approaches did not matter. Existential therapy is distinguished by a sense of reality and concreteness.

I would phrase the above point positively as follows: existential technique should have flexibility and versatility, varying from patient to patient and from one phase to another in treatment with the same patient. The specific technique to be used at a given point should be decided on the basis of these questions: What will best reveal the existence of this particular patient at this moment in his history? What will best illuminate his being in the world? Never merely "eclectic," this flexibility always involves a clear understanding of the underlying assumptions of any method. Let us say a Kinseyite, for example, a traditional Freudian, and an existential analyst are dealing with an instance of sexual repression. The Kinseyite would speak of it in terms of finding a sexual object, in which case he is not talking about sex in human beings. The traditional Freudian would see its psychological implications, but would look primarily for causes in the past and might well ask how this instance of sexual repression *qua* repression can be overcome. The existential therapist would view the sexual repression as a holding back of *potentia* of the existence of this person; and though he might or might not, depending on the circumstances, deal immediately with the sex problem as such, it would always be seen not as a mechanism of repression as such but as a limitation of this person's being in his world.

The second implication is that psychological dynamisms always take their meaning from the existential situation of the patient's own, immediate life. The writings of a number of existential psychotherapists, including van den Berg, Frankl, Boss, and especially the work of R. J. Laing, are pertinent. Some hold that Freud's practice was right but his theories explaining his practice were wrong. Some are Freudian in technique but put the theories and concepts of orthodox psychoanalysis on a fundamental existential basis. Take *transference,* for example, a discovery which many greatly value. What really happens is not that the neurotic patient "transfers" feelings he had toward mother or father to wife or therapist. Rather, the neurotic is one who in certain areas never developed beyond the limited and restricted forms of experience characteristic of the infant. Hence in later years he perceives wife or therapist through the same restricted, distorted "spectacles" as he perceived father or mother. The problem is to be understood in terms of perception and relatedness to the world. This makes unnecessary the concept of transference in the sense of a displacement of detachable feelings from one object to another. The new basis of this concept frees psychoanalysis from the burden of a number of insoluble problems.

Take also the ways of behaving known as *repression* and *resistance.* Freud saw repression as related to bourgeois morality—specifically, as the patient's need to preserve an acceptable picture of himself and, therefore, to hold back thoughts, desires, and so forth which are unacceptable according to bourgeois moral codes. Existential therapists see the conflict more basically in the area of the patient's acceptance or rejection of his own potentialities. We need to keep in mind the question: What keeps the patient from accepting in freedom his potentialities? This may involve bourgeois morality, but it also involves a lot more: it leads immediately to the existential question of the person's freedom. Before repression is possible or conceivable, the person must have some possibility of accepting or rejecting—that is, some margin of freedom. Whether the person is aware of this freedom or can articulate it is another question; he does not need to be. *To repress is precisely to make one's self unaware of freedom.* This is the nature of the dynamism. Thus, to repress or deny this freedom already presupposes it as a possibility. We wish to emphasize that psychic determinism is always a secondary phenomenon and works only in a limited area. The primary question is how the person relates to his freedom to express potentialities in the first place, repression being one way of so relating.

With respect to *resistance,* the question is to be asked: What makes such a phenomenon possible? One answers that it is an outworking of the tendency of the patient to become absorbed in the *Mitwelt,* to slip back into *das Mann,* the anonymous mass, and to renounce the particular unique and original potentiality which is his. Thus "social conformity" is a general form of resistance in life, and even the patient's acceptance of the doctrines and interpretations of the therapist may itself be an expression of resistance.

We do not wish here to go into the question of what underlies these phenomena. We want only to demonstrate that at each point in considering these dynamisms of transference, resistance, and repression, these therapists do something important for the existential approach. *They see each dynamism on an ontological basis.* Each way of behaving is seen and understood in the light of the existence of the patient as a human being. This is shown, too, in the therapist's conceiving of drives, libido, and so forth always in terms of *potentialities* for existence. Boss thereby seeks "to throw overboard the painful intellectual acrobatic of the old psychoanalytic theory which sought to derive the phenomena from the interplay of some forces or drives behind them." He does not deny forces as such but holds that they cannot be understood as "energy transformation" or on any other such natural science model but only as the person's *potentia* of existence. "This freeing from unneccessary constructions facilitates the understanding between patient and doctor. Also it makes the pseudo-resistances disappear which were a justified defense of the analysands against a violation of their essence." Boss holds that he thus can follow the "basic rule" in analysis—the one condition Freud set for analysis, namely, that the patient give forth in complete honesty whatever is going on in his mind—more effectively than in traditional psychoanalysis, for he listens with respect and takes seriously and without reserve the contents of the patient's communication rather than sieving it through prejudgments or destroying it by special interpretations. Boss holds he is simply engaged in bringing out the underlying meaning of Freud's discoveries and placing them on their necessary comprehensive foundation. Believing that Freud's discoveries have to be understood below their faulty formulation, Boss points out that Freud himself was not merely a passive "mirror" for the patient in analysis, as traditionally urged in psychoanalysis, but was "translucent," a vehicle and medium through which the patient sees himself.

The third implication in existential therapy is the emphasis on *presence.* By this we mean that the relationship of the therapist and patient is taken as a real one, the therapist being not merely a shadowy reflector but an alive human being who happens, at that hour, to be concerned not with his own problems but with understanding and experiencing as far as possible the being of the patient. The way was prepared for this emphasis on presence by our discussion above of the fundamental existential idea of truth in relationship. It was there pointed out that existentially truth always involves the relation of the person to something or someone, and that the therapist is part of the patient's relationship "field." We indicated, too, that this was not only the therapist's best avenue to understanding the patient but that he cannot really *see* the patient unless he participates in the field.

Several quotations will make clearer what this presence means. Karl Jaspers has remarked, "What we are missing! What opportunities of understanding we let pass by because at a single decisive moment we were, with all our knowledge, lacking in the simple virtue of a *full human presence!*"[3] In similar vein but greater detail Binswanger writes in his paper on psychotherapy, concerning the significance of the therapist's role of the relationship:

If such a (psychoanalytic) treatment fails, the analyst inclines to assume that the patient is not capable of overcoming his resistance to the physician, for example, as a "father image." Whether an analysis can have success or not is often, however, not decided by whether a patient is capable *at all* of overcoming such a transferred father image but by the opportunity *this particular physician* accords him to do so; it may, in other words, be the rejection of the therapist as a person, the impossibility of entering into a genuine communicative rapport with him, that may form the obstacle against breaking through the "eternal" repetition of the father resistance. Caught in the "mechanism" and thus in what inheres in it, *mechanical repetition,* the psychoanalytic doctrine, as we know, is altogether strangely blind toward the entire category of the *new,* the properly *creative* in the life of the psyche everywhere. Certainly it not always is true to the facts if one attributes the failure of treatment only to the patient; the question always to be asked first by the physician is whether the fault may not be his. What is meant here is not any technical fault but the far more fundamental failure that consists of an impotence to wake or rekindle that divine "spark" in the patient which only true communication from existence to existence can bring forth and which alone possesses, with its light and warmth, also the fundamental power that makes any therapy work—the power to liberate a person from the blind isolation, the *idios kosmos* of Heraclitus, from a mere vegetating in his body, his dreams, his private wishes, his conceit and his presumptions, and to ready him for a life of *koinonia,* of genuine community.[4]

Presence is not to be confused with a sentimental attitude toward the patient but depends firmly and consistently on how the therapist conceives of human beings. It is found in therapists of various schools and differing beliefs—differing, that is, on anything except one central issue: their assumptions about whether the human being is an object to be analyzed or a being to be understood. Any therapist is existential to the extent that, with all his technical training and his knowledge of transference and dynamisms, he is still able to relate to the patient as "one existence communicating with another," to use Binswanger's phrase. In my own experience, Frieda Fromm-Reichmann particularly had this power in a given therapeutic hour; she used to say, "The patient needs an experience, not an explanation." Erich Fromm, for another example, not only emphasizes presence in a way similar to Jasper's statement above but makes it a central point in his teaching of psychoanalysis.

Carl Rogers is an illustration of one who, never having had, so far as I know, direct contact with the existential therapists as such, has written a very existential document in his *apologia pro vita sua* as a therapist:

> I launch myself into the therapeutic relationship having a hypothesis, or a faith, that my liking, my confidence, and my understanding of the other person's inner world, will lead to a significant process of becoming. I enter the relationship not as a scientist, not as a physician who can accurately diagnose and cure, but as a person, entering into a personal relationship. Insofar as I see him only as an object, the client will tend to become only an object.
>
> I risk myself, because if, as the relationship deepens, what develops is a failure, a regression, a repudiation of me and the relationship by the client, then I sense that I will lose myself, or a part of myself. At times this risk is very real, and is very keenly experienced.

I let myself go into the immediacy of the relationship where it is my total organism which takes over and is sensitive to the relationship, not simply my consciousness. I am not consciously responding in a planful or analytic way, but simply in an unreflective way to the other individual, my reaction being based (but not consciously) on my total organismic sensitivity to this other person. I live the relationship on this basis.[5]

There are real differences between Rogers and the existential therapists, such as the fact that most of his work is based on relatively shorter-time therapeutic relationships whereas the work of the existential therapists in this volume is generally long-time. Rogers's viewpoint is at times naïvely optimistic, whereas the existential approach is oriented more to the tragic crises of life, and so forth. What are significant, however, are Rogers's basic ideas that therapy is a "process of becoming," that the freedom and inner growth of the individual are what counts, and the implicit assumption pervading Rogers's work of the dignity of the human being. These concepts are all very close to the existentialist approach to the human being.

Before leaving the topic of *presence,* we need to make three caveats. One is that this emphasis on relationship is in no way an oversimplification or shortcut; it is not a substitute for discipline or thoroughness of training. It rather puts these things in their context—namely, discipline and thoroughness of training directed to understanding human beings as human. The therapist is assumedly an expert. But, if he is not first of all a human being, his expertness will be irrelevant and possibly harmful. The distinctive character of the existential approach is that understanding *being human* is no longer just a "gift," an intuition, or something left to chance. It is the "proper study of man," in Alexander Pope's phrase, and becomes the center of a thorough and scientific concern in the broad sense. The existential analysts do the same thing with the structure of human existence that Freud did with the structure of the unconscious—namely, they take it out of the realm of the hit-and-miss gift of special intuitive individuals, accept it as the area of exploration and understanding, and make it to some extent teachable.

Another caveat is that the emphasis on the reality of presence does not obviate the exceedingly significant truths in Freud's concept of transference, rightly understood. It is demonstrable every day in the week that patients, and all of us to some extent, behave toward therapist or wife or husband as though they were father or mother or someone else, and the working through of this is of crucial importance. But in existential therapy "transference" gets placed in the new context of *an event occurring in a real relationship between two people.* Almost everything the patient does vis-à-vis the therapist in a given hour has an element of transference in it. But nothing is ever "just transference," to be explained to the patient as one would an arithmetic problem. The concept of "transference" as such has often been used as a convenient protective screen behind which both therapist and patient hide in order to avoid the more anxiety-creating situation of direct confrontation. For me to tell myself, say when especially fatigued, that the patient is so demanding because she wants to prove she can make her father love her may be a relief and may also be in fact true. But the real point is that she is doing this to me in this given moment, and the reasons it occurs at this instant of intersection of her existence and mine are not exhausted by what she did with her father. Beyond all considerations of unconscious determinism—which are true in their partial context—she is at some point choosing to do this at this specific moment. Furthermore, the only thing that will grasp the patient, and in the long run make it possible for her to change, is to experience fully and deeply that she is doing precisely this to a real person, myself, in this real moment.[6] Part of the *sense of timing* in therapy, which has received special development among the existential therapists, consists of letting the patient experience what he or she is doing until the experience really grasps him or her.[7] Then and only then will an explanation of *why* he[...] For the patient referred to above to become aware that she is demanding this particular unconditioned love from this real person in this immediate hour may indeed shock her, and thereafter—or possibly only hours later—she should become aware of the early childhood antecedents. She may well explore and re-experience then how she smoldered with anger as a child because she couldn't make her father notice her. But if she is simply told this is a transference phenomenon, she may have learned an interesting intellectual fact which does not existentially grasp her at all.

Another caveat is that *presence* in a session does not at all mean the therapist imposes himself or his ideas or feelings on the patient. It is a highly interesting proof of our point that Rogers, who gives such a vivid picture of presence in the quotation above, is precisely the psychologist who has most unqualifiedly insisted that the therapist not project himself but at every point follow the affect and leads of the patient. Being alive in the relationship does not at all mean the therapist will chatter along with the patient. He will know that patients have an infinite number of ways of trying to become involved with the therapist in order to avoid their own problems. And the therapist may well be silent, aware that to be a projective screen is one aspect of his part of the relationship. The therapist is what Socrates named the "midwife"—completely real in "being there," but being there with the specific purpose of helping the other person to bring to birth something from within himself.

The fourth implication for technique in existential analysis follows immediately from our discussion of presence: therapy will attempt to "analyze out" the ways of behaving which destroy presence. The therapist, on his part, will need to be aware of whatever in him blocks full presence. I do not know the context of Freud's remark that he preferred that patients lie on the couch because he could not stand to be stared at for

.ine hours a day. But it is obviously true that any therapist—whose task is arduous and taxing at best—is tempted at many points to evade the anxiety and potential discomfort of confrontation by various devices. We have earlier described the fact that real confrontation between two people can be profoundly anxiety-creating. Thus it is not surprising that it is much more comfortable to protect ourselves by thinking of the other only as a "patient" or focusing only on certain mechanisms of behavior. The *technical* view of the other person is perhaps the therapist's most handy anxiety-reducing device. This has its legitimate place. The therapist is presumably an expert. But technique must not be used as a way of blocking presence. Whenever the therapist finds himself reacting in a rigid or preformulated way, he had obviously best ask himself whether he is not trying to avoid some anxiety and as a result is losing something existentially real in the relationship. The therapist's situation is like that of the artist who has spent many years of disciplined study learning technique. But he knows that if specific thoughts of technique preoccupy him when he actually is in the process of painting, he has at that moment lost his vision. The creative process, which should absorb him, transcending the subject-object split, has become temporarily broken; he is now dealing with objects and himself as a manipulator of objects.

The fifth implication has to do with the goal of the therapeutic process. The aim of therapy is that the patient *experience his existence as real.* The purpose is that he become aware of his existence as fully as possible, which includes becoming aware of his potentialities and becoming able to act on the basis of them. The characteristic of the neurotic is that his existence has become "darkened," as the existential analysts put it, blurred, easily threatened and clouded over, and gives no sanction to his acts. The task of therapy is to illuminate the existence. The neurotic is overconcerned about the *Umwelt,* and underconcerned about *Eigenwelt.* As the *Eigenwelt* becomes real to him in therapy, the patient tends to experience the *Eigenwelt* of the therapist as stronger than his own. Binswanger points out that the tendency to take over the therapist's *Eigenwelt* must be guarded against, and therapy must not become a power struggle between the two *Eigenwelten.* The therapist's function is to *be there* (with all of the connotation of *Dasein*), present in the relationship, while the patient finds and learns to live out his own *Eigenwelt.*

An experience of my own may serve to illustrate one way of taking the patient existentially. I often have found myself having the impulse to ask, when the patient comes in and sits down, not "*How* are you?" but "*Where* are you?" The contrast of these questions—neither of which would I probably actually ask aloud—highlights what is sought. I want to know, as I experience the patient in this hour, not just how he feels, but rather *where he is,* the "where" including his feelings but also a lot more—whether he is detached or fully present, whether his direction is toward me and toward his problems or away from both, whether he is running from anxiety, whether his special courtesy when he came in or appearance of eagerness to reveal things is really inviting me to overlook some evasion he is about to make, where he is in relation to the girl friend he talked about yesterday, and so on. I became aware of this asking where the patient was before I specifically knew the work of the existential therapists. It illustrates a spontaneous existential attitude.

It follows that when mechanisms or dynamisms are interpreted, as they will be in existential therapy as in any other, it will always be in the context of this person's becoming aware of his existence. This is the only way the dynamism will have reality for the patient, will affect him. Otherwise he might as well—as indeed most patients do these days—read about the mechanism in a book. This point is of special importance because precisely the problem of many patients is that they think and talk about themselves in terms of mechanisms. It is their way, as well-taught citizens of twentieth-century Western culture, to avoid confronting their own existence, their method of repressing ontological awareness. This is done, to be sure, under the rubric of being "objective" about oneself. But is it not, in therapy as well as in life, often a systematized, culturally acceptable way of rationalizing detachment from oneself? Even the motive for coming for therapy may be just that—to find an acceptable system by which one can continue to think of oneself as a mechanism, to run oneself as one would a motor car, only now to do it successfully. If we assume, as we have reason for doing, that the fundamental neurotic process in our day is the repression of the ontological sense—the loss of the sense of being, together with the truncation of awareness and the locking up of the potentialities which are the manifestations of this being—then we are playing directly into the patient's neurosis to the extent that we teach him new ways of thinking of himself as a mechanism. This is one illustration of how psychotherapy can reflect the fragmentation of the culture, structuralizing neurosis rather than curing it. Trying to help the patient with a sexual problem by explaining it merely as a mechanism is like teaching a farmer irrigation while damming up his streams of water.

This raises some penetrating questions about the nature of "cure" in psychotherapy. It implies that it is not the therapist's function to "cure" the patients' neurotic symptoms, though this is the motive for which most people come for therapy. Indeed, the fact that this is their motive reflects their problem. Therapy is concerned with something more fundamental—namely, helping the person experience his existence—and any cure of symptoms which will last must be a by-product of that. The general idea of "cure"—namely, to become as satisfactorily adjusted as possible—is itself a denial of *Dasein,* of this particular patient's being. The kind of cure that consists of adjustment, becoming able to fit the culture, can be obtained by technical emphases in therapy, for it is precisely the central theme of the culture that one live in a calculated, controlled, technically well-managed way. Then the patient accepts a confined world without conflict, for now his world is identical with the culture. And since anxiety comes only with freedom, the patient naturally gets over his anxiety; he is relieved from his symptoms because he surrenders the possibilities which caused his anxiety. This is the way of being "cured" by giving up being, giving up existence, by constricting, hedging in existence. In this respect, psychotherapists become the agents of the culture whose particular task it is to adjust people to it; psychotherapy becomes an expression of the fragmentation of the period rather than an enterprise for overcoming it. As we have indicated above, there are clear historical indications that this is occurring in the different psychotherapeutic schools, and the historical probability is that it will increase. There is certainly a question how far this gaining of release from conflict by giving up one's being can proceed without generating in individuals and groups a submerged despair, a resentment which will later burst out in self-destructiveness, for history proclaims again and again that sooner or later man's need to be free will out. But the complicating factor in our immediate historical situation is that the culture itself is built around this ideal of technical adjustment and carries so many built-in devices for narcotizing the despair that come from using oneself as a machine that the damaging effects may remain submerged for some time.

On the other hand, the term *cure* can be given a deeper and truer meaning—namely, becoming oriented toward the fulfillment of one's existence. This may well include as a by-product the cure of symptoms—obviously a desideratum, even if we have stated decisively that it is not the chief goal of therapy. The important thing is that the person discovers his being, his *Dasein*.

The sixth implication which distinguishes the process of existential therapy is the importance of *commitment*. The basis for this was prepared at numerous points in our previous sections, particularly in our discussion of Kierkegaard's idea that "truth exists only as the individual himself produces it in action." The significance of commitment is not that it is simply a vaguely good thing or ethically to be advised. It is a necessary prerequisite for seeing truth. This involves a crucial point which has never to my knowledge been fully taken into account in writings on psychotherapy—namely, that *decision precedes knowledge.* We have worked normally on the assumption that, as the patient gets more and more knowledge and insight about himself, he will make the appropriate decisions. This is a half truth. The second half of the truth is generally overlooked—namely, that *the patient cannot permit himself to get insight or knowledge until he is ready to decide, until he takes a decisive orientation to life and has made the preliminary decisions along the way.*

We mean "decision" here not in the sense of a be-all-and-end-all jump—say, to get married or to join the foreign legion. The possibility or readiness to take such "leaps" is a necessary condition for the decisive orientation, but the big leap itself is sound only so far as it is based upon the minute decisions along the way. Otherwise the sudden decision is the product of unconscious processes, proceeding compulsively in unawareness to the point where they erupt—for example, in a "conversion." We use the term *decision* as meaning a decisive attitude toward existence," an attitude of commitment. In this respect, *knowledge and insight follow decision rather than vice versa.* Everyone knows of the incidents in which a patient becomes aware in a dream that a certain boss is exploiting him and the next day decides to quit his job. But just as significant, though not generally taken into account because they go against our usual ideas of causality, are the incidents when the patient cannot have the dream *until* he makes the decision. He makes the jump to quit his job, for example, and then he can permit himself to see in dreams that his boss was exploiting him all along.

One interesting corollary of this point is seen when we note that a patient cannot recall what was vital and significant in his past until he is ready to make a decision with regard to the future. Memory works not on a basis simply of what is there imprinted; it works on the basis of one's decisions in the present and future. It has often been said that one's past determines one's present and future. Let it be underlined that one's present and future—how he commits himself to existence at the moment—also determines his past. That is, it determines what he can recall of his past, what portions of his past he selects (consciously but also unconsciously) to influence him now, and therefore the particular Gestalt his past will assume.

This commitment is, furthermore, not a purely conscious or voluntaristic phenomenon. It is also present on so-called "unconscious" levels. When a person lacks commitment, for example, his dreams may be staid, flat, impoverished. But when he does assume a decisive orientation toward himself and his life, his dreams often take over the creative process of exploring, molding, forming himself in relation to his future, or—what is the same thing from the neurotic viewpoint—the dreams struggle to evade, substitute, cover up. The important point is that either way the issue has been joined.

With respect to helping the patient develop the orientation of commitment, we should first emphasize that the existential therapists do not at all mean activism. This is no "decision as a short cut," no matter of premature jumping because to act may be easier and may quiet anxiety more quickly than the slow, arduous, long-time process of self-exploration. They mean rather the attitude of *Dasein,* the self-aware being taking his own existence seriously. The points of commitment and decision are those where the dichotomy between being subject and object is overcome in the unity of readiness for action. When a patient discusses intellectually *ad interminum* a given topic without its ever shaking him or becoming real to him, the therapist asks what is he doing existentially by means of this talk? The talk itself, may well be in the service of covering up reality, rationalized generally under the idea of unprejudiced inquiry into the data. It is customarily said that the patient will break through such talk when some experience of anxiety, some inner suffering or outer threat, shocks him into committing himself really to getting help and gives him the incentive necessary for the painful process of uncovering illusions, of inner change and growth. True, this does occur from time to time. And the existential therapist can aid the patient in absorbing the real impact of such experiences by helping him develop the capacity for silence (which is another form of communication) and thus avoid using chatter to break the shocking power of the encounter with the insight.

But in principle I do not think the conclusion that we must wait around until anxiety is aroused is adequate. If we assume that the patient's commitment depends upon being pushed by external or internal pain, we are in several difficult dilemmas. Either the therapy "marks time" until anxiety or pain occurs, or we arouse anxiety ourselves (which is a questionable procedure). And the very reassurance and quieting of anxiety the patient receives in therapy may work against his commitment to further help and may make for postponement and procrastination.

Commitment must be on a more positive basis. The question we need to ask is: What is going on that the patient has not found some point in his own existence to which he can commit himself unconditionally? In the earlier discussion of nonbeing and death, it was pointed out that everyone constantly faces the threat of nonbeing if he lets himself recognize the fact. Central here is the symbol of death, but such threat of destruction of being is present in a thousand and one other guises as well. The therapist is doing the patient a disservice if he takes away from him the realization that it is entirely within the realm of possibility that he forfeit or lose his existence and that may well be precisely what he is doing at this very moment. This point is especially important because patients tend to carry a never-quite-articulated belief, no doubt connected with childhood omnipotent beliefs associated with parents, that somehow the therapist will see that nothing harmful happens to them and therefore they don't need to take their own existence seriously. The tendency prevails in much therapy to water down anxiety, despair, and the tragic aspects of life. Is it not true as a general principle that we need to engender anxiety only to the extent that we already have watered it down? Life itself produces enough, and the only real, crises; and it is very much to the credit of the existential emphasis in therapy that it confronts these tragic realities directly. The patient can indeed destroy himself if he so chooses. The therapist may not say this: it is simply a reflection of fact, and the important point is that it not be sloughed over. The symbol of suicide as a possibility has a far-reaching positive value. Nietzsche once remarked that the thought of suicide has saved many lives. I am doubtful whether anyone takes his life with full seriousness until he realizes that it is entirely within his power to commit suicide.[8]

Death in any of its aspects is the fact which makes of the present hour something of absolute value. One student put it, "I know only two things—one, that I will be dead someday, two, that I am not dead now. The only question is what shall I do between those two points." We cannot go into this matter in further detail, but we only wish to emphasize that the core of the existential approach is the taking of existence seriously.

We conclude with two final caveats. One is a danger that lies in the existential approach, the danger of *generality.* It would indeed be a pity if the existential concepts were tossed around among therapists without regard for their concrete, real meaning. For it must be admitted that there is temptation to become lost in words in these complex areas with which existential analysis deals. One can certainly become philosophically detached in the same way as one can be technically detached. The temptation to use existential concepts in the service of intellectualizing tendencies is especially to be guarded against since, because they refer to things that have to do with the center of personal reality, these concepts can the more seductively give the illusion of dealing with reality. Some readers may feel that I have not fully resisted this temptation. I could plead the necessity of having to explain a great deal within a short compass, but extenuating circumstances are not the point. The point is that to the extent that the existential movement in psychotherapy becomes influential in this country—a desideratum which we believe would be very beneficial—the adherents will have to be on guard against the use of the concepts in the service of intellectual detachment. It is precisely for the above reasons that the existential therapists pay much attention to making clear the verbal utterances of the patient, and they also continually make certain that the necessary interrelation of verbalizing and acting is never overlooked. The "logos must be made flesh." The important thing is *to be* existential.

The other caveat has to do with the existential attitude toward the *unconscious.* In principle most existential analysts deny this concept. They point out all the logical as well as psychological difficulties with the doctrine of the unconscious, and they stand against splitting the being into parts. What is called unconscious, they hold, is still part of this given person; *being,* in any living sense, is at its core indivisible.

Now it must be admitted that the doctrine of the unconscious has played most notoriously into the contemporary tendencies to rationalize behavior, to avoid the reality of one's own existence, to act as though one were not himself doing the living. (The man in the street who has picked up the lingo says, "My unconscious did it.") The existential analysts are correct, in my judgment, in their criticism of the doctrine of the unconscious as a convenient blank check on which any causal explanation can be written or as a reservoir from which any deterministic theory can be drawn. But this is the "cellar" view of the unconscious, and objections to it should not be permitted to cancel out the great contribution that the historical meaning of the unconscious had in Freud's terms. Freud's great discovery and his perdurable contribution was to enlarge the sphere of the human personality beyond the immediate voluntarism and rationalism of Victorian man to include in this enlarged sphere the "depths"—that is, the irrational, the so-called repressed, hostile, and unacceptable urges, the forgotten aspects of experience, *ad infinitum.* The symbol for this vast enlarging of the domain of the personality was "the unconscious."

I do not wish to enter into the complex discussion of this concept of the unconscious itself. I wish only to suggest a position. It is right that the blank-check, deteriorated, cellar form of this concept should be rejected. But the far-reaching enlargement of personality, which is its real meaning of the concept of the unconscious, should not be lost. Binswanger remarks that, for the time being, the existential therapists will not be able to dispense with the concept of the unconscious. I would propose, rather, that being is indivisible, that unconsciousness is part of any given being, that the cellar theory of the unconscious is logically wrong and practically unconstructive, but that the meaning of the discovery—namely, the radical enlargement of being—is one of the great contributions of our day and must be retained.

Notes

TWO The Case of Mrs. Hutchens

Adapted from *American Journal of Orthopsychiatry* 30 (1960):685–695.

1. From a philosophical point of view, these would be termed "ontological principles."

THREE Origins and Significance of Existential Psychology

1. L. Binswanger, "Existential Analysis and Psychotherapy," in *Progress in Psychotherapy,* ed. Fromm-Reichmann and Moreno (New York: Grune & Stratton, 1956), p. 144.

2. Personal communication from Dr. Lefebre, an existential psychotherapist who was a student of Jaspers and Boss.

3. Binswanger, p. 145.

4. L. Binswanger, "The Case of Ellen West," in *Existence: A New Dimension in Psychology and Psychiatry,* ed. Rollo May, Ernest Angel, and Henri Ellenberger (New York: Basic Books, 1958), pp. 237–364.

5. Sigmund Freud *Introductory Lectures on Psychoanalysis,* trans. and ed. James Strachey (New York: Liveright, 1979).

6. Binswanger, "The Case of Ellen West," p. 294.

7. L. Binswanger, *Sigmund Freud: Reminiscences of a Friendship,* trans. Norbert Guterman (New York: Grune and Stratton, 1957).

8. Helen Sargent, "Methodological Problems in the Assessment of Intrapsychic Change in Psychotherapy," unpublished paper.

9. *Existence,* pp. 92–127.

10. Gordon Allport, *Becoming, Basic Considerations for a Psychology of Personality* (New Haven: Yale University Press, 1955).

11. To see this one has only to name the originators of new theory: Freud, Adler, Jung, Rank, Stekel, Reich, Horney, Fromm, etc. The two exceptions, so far as I can see, are the schools of Harry Stack Sullivan and Carl Rogers, and the former was indirectly related to the work of the Swiss-born Adolph Meyer. Even Rogers may partly illustrate our point, for although his approach has clear and consistent theoretical implications about human nature, his focus has been on the "applied" rather than the "pure" science side, and his theory about human nature owes much to Otto Rank.

12. Quoted by Paul Tillich, "Existential Philosophy," *Journal of the History of Ideas* 5 (1944):44–70.

13. John Wild, *The Challenge of Existentialism* (Bloomington: Indiana University Press, 1955). Modern physics, with Heisenberg, Bohr, and similar trends have changed at this point, paralleling, as we shall see later, one side of the existentialist development. We are talking above of the traditional ideas of Western science.

14. Kenneth W. Spence, *Behavior Theory and Conditioning* (New Haven: Yale University Press, 1956).

15. Tillich, *op. cit.*

16. Those who wish to know more about the existential movement as such are referred to Tillich, *op. cit.* For most of the above historical material I am indebted to Tillich's paper.

17. Martin Heidegger, *Being and Time,* trans. John Macquarrie and Edward Robinson (New York: Harper & Row, 1962). Heidegger disclaimed the title "existentialist" after it became identified with the work of Sartre. He would call himself, strictly speaking, a philologist or ontologist. But in any case, we must be existential enough not to get twisted up in controversies over titles and to take the meaning and spirit of each man's work rather than the letter. Martin Buber likewise is not happy at being called an existentialist, although his work has clear affinities with this movement. The reader who has difficulty with the terms in this field is indeed in good company!

18. Paul Tillich, *The Courage to Be* (New Haven: Yale University Press, 1952) is existential as a living approach to crises in contrast to books *about* existentialism. Tillich, like most of the thinkers mentioned above, is not to be tagged as *merely* an existentialist, for existentialism is a way of approaching problems and does not in itself give answers or norms. Tillich has both rational norms—the structure of reason is always prominent in his analyses—and religious norms. Some readers will not find themselves in agreement with the religious elements in *The Courage to Be.* It is important to note the very significant point, however, that these religious ideas, whether one agrees with them or not, do illustrate an authentic existential approach. This is seen in Tillich's concept of "the God beyond God" and "absolute faith" as faith not *in* some content or somebody but as a state of being, a way of relating to reality characterized by courage, acceptance, full commitment, etc. The theistic arguments for the "existence of God" are not only beside the point but exemplify the most deteriorated aspect of the Western habit of thinking in terms of God as a substance or object, existing in a world of objects and in relation to whom we are subjects. This is "bad theology," Tillich points out, and results in "the God Nietzsche said had to be killed because nobody can tolerate being made into a mere object of absolute knowledge and absolute control" (p. 185).

19. Paul Tillich, "Existential Aspects of Modern Art," in *Christianity and the Existentialists,* ed. Carl Michalson (New York: Scribners, 1956), p. 138.

20. José Ortega y Gasset, *The Dehumanization of Art, and Other Writings on Art and Culture* (New York: Doubleday Anchor, 1956), pp. 135–137.

21. Pascal, *Pensées,* ed. and trans. Gertrude B. Burford Rawlings (New York: Peter Pauper Press, 1946), p. 36.

22. It is not surprising, thus, that this approach to life would speak particularly to many modern citizens who are aware of the emotional and spiritual dilemmas in which we find ourselves. Norbert Wiener, for example, though the actual implications of his scientific work may be radically different from the emphases of the existentialists, stated in his autobiography that his scientific activity led him personally to a "positive" existentialism. "We are not fighting for a definitive victory in the indefinite future," he writes. "It is the greatest possible victory *to be,* and *to have been.* No defeat can deprive us of the success of having existed for some moment of time in a universe that seems indifferent to us." *I Am a Mathematician* (New York: Doubleday). (Italics mine.)

23. Witter Bynner, *The Way of Life, According to Laotzu, an American Version* (New York: John Day, 1946).

24. See William Barrett, ed., *Zen Buddhism, the Selected Writings of D. T. Suzuki* (New York: Doubleday Anchor, 1956), p. xi.

FOUR How Existentialism and Psychoanalysis Arose Out of the Same Cultural Situation

1. SöreN Kierkegaard, *The Sickness unto Death,* trans. Walter Lowrie (New York: Doubleday, 1954).

2. Ernest Schachtel, "On Affect, Anxiety and the Pleasure Principle," in *Metamorphosis* (New York: Basic Books, 1959), pgs. 1–69.

3. Ernest Cassirer, *An Essay on Man* (New Haven: Yale University Press, 1944), p. 21.

4. Max Scheler, *Die Stellung des Menschen im Kosmos* (Darmstadt: Reichl, 1928), pp. 13 f.

5. Sigmund Freud, *Civilization and Its Discontents,* trans. and ed. James Strachey (New York: Norton, 1962).

6. Walter A. Kaufmann, *Nietzsche: Philosopher, Psychologist, AntiChrist* (Princeton: Princeton University Press, 1950), p. 140.

FIVE Kierkegaard, Nietzsche, and Freud

1. Rollo May, *The Meaning of Anxiety,* rev. ed. (New York: Norton, 1977), pp. 36–52. Those pages may be recommended as a short survey of the importance of Kierkegaard's ideas. His two most important psychological books are *The Concept of Anxiety* and *The Sickness unto Death.* For further acquaintance with Kierkegaard, see *A Kierkegaard Anthology,* ed. Robert Bretall (Princeton: Princeton University Press, 1951).

2. Walter A. Kaufmann, *Nietzsche: Philosopher, Psychologist, AntiChrist* (Princeton: Princeton University Press, 1950), p. 135.

3. Thus the very increase of truth may leave human beings less secure, if they let the objective increase of truth act as a substitute for their own commitment, their own

relating to the truth in their own experience. He "who has observed the contemporary generation," wrote Kierkegaard, "will surely not deny that the incongruity in it and the reason for its anxiety and restlessness is this, that in one direction truth increases in extent, in mass, partly also in abstract clarity, whereas certitude steadily decreases."

4. See Walter Lowrie, *A Short Life of Kierkegaard* (Princeton: Princeton University Press, 1942).

5. Quoted from the "Concluding Unscientific Postscript," in *A Kierkegaard Anthology, op. cit.*, pp. 210–211. (Kierkegaard has the whole passage in italics; we have limited them, for purposes of contrast, to the new element—namely, the subjective relation to truth.) It is highly interesting that the example Kierkegaard goes on to cite, after the above sentences, is the knowledge of God, and points out—a consideration that would have saved endless confusions and futile bickerings—that the endeavor to prove God as an "object" is entirely fruitless, and that truth rather lies in the nature of the relationship ("even if he should happen to be thus related to what is not true"). It should be evident that Kierkegaard is not implying that whether or not something is objectively true doesn't matter. That would be absurd. He is referring, as he phrases it in a footnote, to "the truth which is essentially related to existence."

6. See Martin Heidegger, "On the Essence of Truth," in *Existence and Being,* ed. Werner Brock (South Bend, Ind.: Regnery, 1949).

7. From mimeographed address by Werner Heisenberg, Washington University, St. Louis, Oct. 1954.

8. It should be possible to demonstrate—possibly it has already been done—in perception experiments that the interest and involvement of the observer increase the accuracy of his perception. There are indications already in Rorschach responses that in the cards where the subject becomes emotionally involved, his perception of form becomes more, not less, sharp and accurate. (I am speaking not of neurotic emotion; that introduces different factors.)

9. Both Kierkegaard and Nietzsche knew that "man cannot sink back into unreflective immediacy without losing himself; but he can go this way to the end, not destroying reflection, but rather coming to the basis in himself in which reflection is rooted." Thus speaks Karl Jaspers in his enlightening discussion of the similarities of Nietzsche and Kierkegaard, whom he regards as the two greatest figures of the nineteenth century. See his book, *Reason and Existence,* trans. William Earle (The Noonday Press, 1955), chap. 1.

10. The existential thinkers as a whole take this loss of consciousness as the centrally tragic problem of our day, not at all to be limited to the psychological context of neurosis. Jaspers indeed believes that the forces which destroy personal consciousness in our time, the juggernaut processes of conformity and collectivism, may well lead to a more radical loss of individual consciousness on the part of modern man.

11. Both Kierkegaard and Nietzsche also share the dubious honor of being dismissed in some allegedly scientific circles as pathological! I assume this fruitless issue needs no longer to be discussed; Binswanger quotes Marcel in a paper concerning those who dismiss Nietzsche because of his ultimate psychosis, "One is free to learn nothing if one wishes." A more fruitful line of inquiry, if we wish to consider the psychological crises of Kierkegaard and Nietzsche, is to ask whether any human being can support

an intensity of self-consciousness beyond a certain point, and whether the creativity (which is one manifestation of this self-consciousness) is not paid for by psychological upheaval.

12. Kaufmann, *op. cit.*, p. 93.
13. *Ibid.*, pp. 133–134.
14. *Ibid.*, p. 229.
15. *Ibid.*, p. 168.
16. *Ibid* p. 239.
17. *Ibid.*, p. 169.
18. *Ibid.*, p. 136.
19. Friedrich Nietzsche, *Genealogy of Morals* (Garden City, N.Y.: Doubleday, 1956), p. 217.
20. *Ibid.*, p. 102.
21. *Ibid.*, p. 247.
22. Ernest Jones, *The Life and Work of Sigmund Freud* (New York: Basic Books, 1955), II, p. 344. Dr. Ellenberger, commenting on the affinities of Nietzsche with psychoanalysis, adds, "In fact, the analogies are so striking that I can hardly believe that Freud never read him, as he contended. Eitehr he must have forgotten that he read him, or perhaps he must have read him in indirect form. Nietzsche was so much discussed everywhere at that time, quoted thousands of times in books, magazines, newspapers, and in conversations in everyday life, that it is almost impossible that Freud could not have absorbed his thought in one way or another." Whatever one may assume at this point, Freud did read Edward von Hartmann (Kris points out), who wrote a book, *The Philosophy of the Unconscious.* Both von Hartmann and Nietzsche got their ideas of the unconscious from Schopenhauer, most of whose work also falls in the existential line.
23. Jones, II, p. 344.
24. *Ibid.*, I, p. 295.
25. *Ibid.*, II, p. 432.
26. Ludwig Binswanger, "The Existential Analysis School of Thought," in *Existence: A New Dimension in Psychology and Psychiatry,* ed. Rollo May, Ernest Angel, and Henri Ellenberger (New York: Basic Books, 1958), pp. 191–213. The point that Freud deals with *homo natura* was centrally made by Binswanger in the address he was invited to give in Vienna on the occasion of the eightieth birthday of Freud.

SIX To Be and Not to Be

1. Jean-Paul Sartre, *Being and Nothingness,* trans. Hazel Barnes (New York: Philosophical Library, 1956), p. 561. Sartre goes on, "either in looking for the *person* we encounter a useless, contradictory metaphysical substance—or else the being whom we seek vanishes in a dust of phenomena bound together by external connections. But what each of us requires in this very effort to comprehend another is that he should never resort to this idea of substance, which is inhuman because it is well this side of the human" (p. 52). Also, "If we admit that the person is a totality, we can not hope

to reconstruct him by an addition or by an organization of the diverse tendencies which we have empirically discovered in him." "Every attitude of the person contains some reflection of this totality," holds Sartre. "A jealousy of a particular date in which a subject posits himself in history in relation to a certain woman, signifies for the one who knows how to interpret it, the total relation to the world by which the subject constitutes himself as a self. In other words this *empirical attitude* is by itself the expression of the 'choice of an intelligible character.' There is no mystery about this" (p. 58).

2. Gabriel Marcel, *The Philosophy of Existence* (1949), p. 1.

3. *Ibid.* Italics mine. For data concerning the "morbid effects of the repression" of the sense of being, cf. Fromm, *Escape from Freedom,* and David Riesman, *The Lonely Crowd.*

4. Marcel, p. 5.

5. Pascal, *Pensées,* ed. and trans. Gertrude B. Burfurd Rawlings (New York: Peter Pauper Press, 1946), p. 35. Pascal goes on: "Thus all our dignity lies in thought. By thought we must raise ourselves, not by space and time, which we cannot fill. Let us strive, then, to think well,—therein is the principle of morality." It is perhaps well to remark that by "thought" he means not intellectualism nor technical reason but self-consciousness, the reason which also knows the reasons of the heart.

6. Since our purpose is merely to illustrate one phenomenon—namely, the experience of the sense of being—we shall not report the diagnostic or other details of the case.

7. Some readers will be reminded of the passage in Exodus 3:14 in which Moses, after Yahweh had appeared to him in the burning bush and charged him to free the Israelites from Egypt, demands that the God tell his name. Yahweh gives the famous answer, "I am that I am." This classical, existential sentence (the patient, incidentally, did not consciously know this sentence) carries great symbolic power because, coming from an archaic period, it has God state that *the quintessence of divinity is the power to be.* We are unable to go into the many rich meanings of this answer, nor the equally intricate translation problems, beyond pointing out that the Hebrew of the sentence can be translated as well, "I shall be what I shall be." This bears out our statement above that being is in the future tense and inseparable from becoming; God is creative *potentia,* the essence of the power to become.

8. I omit for purposes of the above discussion the question whether this rightly should be called "transference" or simply human trust at this particular point in this case. I do not deny the validity of the concept of transference rightly defined, but it never makes sense to speak of something as "just transference," as though it were all simply carried over from the past.

9. William Healy, Agusta F. Bronner, and Anna Mae Bowers, *The Meaning and Structure of Psychoanalysis* (New York: Knopf, 1930), p. 38. We give these quotations from a standard summary from the classical middle period of psychoanalysis, not because we are not aware of refinements made to ego theory later, but because we wish to show the essence of the concept of the ego, an essence which has been elaborated but not basically changed.

10. *Ibid.,* p. 41.

11. *Ibid.*, p. 38.

12. If the objection is entered that the concept of the "ego" at least is more precise and therefore more satisfactory scientifically than this sense of being, we can only repeat what we have said above, that precision can be gained easily enough on paper. But the question always is the bridge between the concept and the reality of the person, and the scientific challenge is to find a concept, a way of understanding, which does not do violence to reality, even though it may be less precise.

13. This is an interpretation of Heidegger, given by Werner Brock in the introduction to *Existence and Being* (South Bend, Ind.: Regnery, 1949), p. 77. For those who are interested in the logical aspects of the problem of being versus nonbeing, it may be added that the dialectic of "yes versus no," as Tillich points out in *The Courage to Be* (New Haven: Yale University Press, 1952), is present in various forms throughout the history of thought. Hegel held that nonbeing was an integral part of being, specifically in the "antithesis" stage of his dialectic of "thesis, antithesis, and synthesis." The emphasis on "will" in Schelling, Schopenhauer, Nietzsche, and others as a basic ontological category is a way of showing that being has the power of "negating itself without losing itself." Tillich, giving his own conclusion, holds that the question of how being and nonbeing are related can be answered only metaphorically: "Being embraces both itself and non-being." In everyday terms, being embraces nonbeing in the sense that we can be aware of death, can accept it, can even invite it in suicide —in short, can by self-awareness encompass death.

SEVEN Anxiety and Guilt as Ontological

1. The points in this summary of ontological anxiety are given in epigrammatic form, since for reasons of space we are forced to omit the considerable empirical data which could be cited at each point. A fuller development of some aspects of this approach to anxiety will be found in my book, *The Meaning of Anxiety*, rev. ed. (New York: Norton, 1977).

2. We speak here of anxiety as the "subjective" state, making a distinction between subjective and objective that may not be entirely justified logically but shows the viewpoint from which one observes. The "objective" side of the anxiety experience, which we can observe from the outside, shows itself in severe cases in disordered, catastrophic behavior (Goldstein) or in cases of neurotics in symptom-formation or in cases of "normal" persons in ennui, compulsive activity, meaningless diversions, and truncation of awareness.

3. See the discussion of this phenomenon in Eugene Minkowski, "Findings in a Case of Schizophrenic Depression," in *Existence: A New Dimension in Psychology and Psychiatry*, ed. Rollo May, Ernest Angel, and Henri Ellenberger (New York: Basic Books, 1958).

4. It is an interesting question whether our pragmatic tendencies in English-speaking countries to avoid reacting to anxiety experiences—by being stoical in Britain and by not crying or showing fear in this country, for examples—is part of the reason we have not developed words to do justice to the experience.

5. May, p. 32.

6. Kurt Goldstein, *Human Nature in the Light of Psychopathology* (Cambridge: Harvard University Press, 1940).

7. There is a legitimate argument that what I call "ontological guilt" ought to be called "existentially universal guilt." The terms mean very much the same thing; hence I decided to leave the above term *ontological* as I originally wrote it. (R.M., 1983.)

8. Medard Boss, *Psychoanalyse und Daseinsanalytik* (Bern and Stuttgart: Verlag Hans Huber, 1957). I am grateful to Dr. Erich Heydt, student and colleague of Boss, for translating parts of this work for me as well as discussing at length with me the viewpoint of Boss.

EIGHT Being in the World

1. L. Binswanger, in *Existence: A New Dimension in Psychology and Psychiatry*, ed. Rollo May, Ernest Angel, and Henri Ellenberger (New York: Basic Books, 1958), p. 197.

2. Note how close this description, amounting almost to a word-for-word prediction, is of what is described by the contemporary psychoanalysts, writing in the mid-seventies, as the "narcissistic personality," especially Heinz Kohut and Otto Kernberg. (R.M., 1983.)

3. This phrase "epistemological loneliness" is used by David Bakan to describe Western man's experience of isolation from his world. He sees this isolation as stemming from the skepticism which we inherited from the British empiricists Locke, Berkeley, and Hume. Their error specifically, he holds, was in conceiving of the "thinker as essentially alone rather than as a member and participant of a thinking community." ("Clinical Psychology and Logic," *American Psychologist* [December 1956]:656). It is interesting that Bakan, in good psychological tradition, interprets the error as a social one—namely, separation from the community. But is this not more symptom than cause? More accurately stated, is not the isolation from the community simply one of the ways in which a more basic and comprehensive isolation shows itself?

4. In *Existence*, p. 142.

5. Thus Heidegger uses the terms *to sojourn* and *to dwell* rather than *is* when he speaks of a person being some place. His use of the term *world* is in the sense of the Greek *kosmos*—that is, the "uni-verse" with which we act and react. He chides Descartes for being so concerned with *res extensa* that he analyzed all the objects and things in the world and forgot about the most significant fact of all—namely, that there is world *itself*, that is, a meaningful relationship of these objects with the person. Modern thought has followed Descartes almost exclusively at this point, greatly to the impairment of our understanding of human beings.

6. See L. Binswanger, *Existence*, p. 196.

7. The term *culture* is generally in common parlance set over against the individual —e.g., "the influence of the culture on the individual." This usage is probably an unavoidable result of the dichotomy between subject and object in which the concepts of "individual" and "culture" emerged. It omits the very significant fact that the individual is at every moment also forming his culture.

8. "World openness is the distinctively human characteristic of man's awake life," Schachtel continues. He discusses cogently and clearly the life space and life time which characterize the human being's world in contrast to that of plants and animals. "In the animals, drives and affects remain to a very large extent ties to an inherited instinctive organization. The animal is embedded in this organization and in the closed world (J. v. Uexküll's 'Werkwelt' and 'Wirkwelt') corresponding to this organization. Man's relation to his world is an open one, governed only to a very small extent by instinctive organization, and to the largest extent by man's learning and exploration, in which he establishes his complex, changing and developing relations with his fellow men and with the natural and cultural world around him." So closely interrelated are man and his world, Schachtel demonstrates, that "all our affects arise from . . . spatial and temporal gaps which open between us and our world." "On Affect, Anxiety and the Pleasure Principle," in *Metamorphosis* (New York: Basic Books, 1959), pp. 19–77.

9. L. Binswanger, "The Existential Analysis School of Thought," in *Existence,* p. 191. In this chapter, it is significant to note the parallels Binswanger draws between his conception of "world" and that of Kurt Goldstein.

10. Roland Kuhn, in *Existence,* pp. 365–425.

NINE The Three Modes of World

1. In this respect it is significant to note that Kierkegaard and Nietzsche, in contrast to the great bulk of nineteenth-century thinkers, were able to take the body seriously. The reason was that they saw it not as a collection of abstracted substances or drives, *but as one mode of the reality of the person.* Thus when Nietzsche says "We think with our bodies," he means something radically different from the behaviorists.

2. Martin Buber has developed implications of *Mitwelt* in his *I and Thou* philosophy. See his lectures at the Washington School of Psychiatry, printed in *Psychiatry* 20 (May 1957), especially the lecture on "Distance and Relation."

3. This concept was originally formulated by William James as "the self is the sum of the different roles the person plays." Though the definition was a gain in its day in overcoming a fictitious "self" existing in a vacuum, we wish to point out that it is an inadequate and faulty definition. If one takes it consistently, one not only has a picture of an *unintegrated,* "neurotic" self but falls into all kinds of difficulty in adding up these roles. I propose, rather, that the self is not the sum of the roles you play but your capacity *to know that you are the one playing these roles.* This is the only point of integration, and rightly makes the roles *manifestations* of the self.

4. Sören Kierkegaard, *Fear and Trembling,* trans. Walter Lowrie (New York: Doubleday, 1954), p. 55.

TEN Of Time and History

1. Eugene Minkowski, "Findings in a Case of Schizophrenic Depression," in *Existence: A New Dimension in Psychology and Psychiatry,* ed. Rollo May, Ernest Angel, and Henri Ellenberger (New York: Basic Books, 1958), pp. 127–139.

2. This understanding of time is also reflected in "process philosophies," such as Whitehead's, and has obvious parallels in modern physics.

3. Cf. Tillich, "Existence is distinguished from essence by its temporal character." Also Heidegger, referring to one's awareness of his own existence in time, "Temporality is the genuine meaning of Care." Paul Tillich, "Existential Philosophy," *Journal of the History of Ideas* 5 (1944):61, 62.

4. O. Hobart Mowrer, "Time as a Determinant in Integrative Learning," in *Learning Theory and Personality Dynamics* (New York: Ronald Press, 1950).

5. Henri Bergson, quoted by Tillich, p. 56.

6. Heidegger's *Being and Time* is devoted, as its title indicates, to an analysis of time in human existence. His overall theme is "the vindication of time for being." He calls the three modes of time—namely, past, present, and future—the "three ectasies of time," using the term *ecstasy* in its etymological meaning of "to stand outside and beyond." For the essential characteristic of the human being is the capacity to transcend a given mode of time. Heidegger holds that our preoccupation with objective time is really an evasion; people much prefer to see themselves in terms of objective time, the time of statistics, of quantitative measurement, of "the average," etc., because they are afraid to grasp their existence directly. He holds, moreover, that objective time, which has its rightful place in quantitative measurements, can be understood only on the basis of time as immediately experienced rather than vice versa.

7. Not only the existential psychologists and psychiatrists but the existential thinkers in general are to be distinguished precisely by the fact that they do take seriously the historical cultural situation which conditions the psychological and spiritual problems for any individual. But they emphasize that to know history we must act in it. Cf. Heidegger: "Fundamentally history takes its start not from the 'present' nor from what is 'real' only today, but from the future. The 'selection' of what is to be an object of history is made by the actual, 'existential' choice . . . of the historian, in which history arises." Martin Heidegger, *Existence and Being,* ed. Werner Brock (South Bend, Ind.: Regnery, 1949), p. 110. The parallel in therapy is that what the patient selects from the past is determined by what he faces in the future.

8. Sören Kierkegaard, *Fear and Trembling,* trans. Walter Lowrie (New York: Doubleday, 1954), p. 130. What we do learn from previous generations are facts; one may learn them by repetition, like the multiplication table, or remember facts or experiences on their "shock" basis. Kierkegaard is not denying any of this. He was well aware that there is progress from one generation to the next in *technical areas.* What he is speaking of is "that which is genuinely human"—specifically, love.

ELEVEN Transcending the Immediate Situation

1. This antagonism was illustrated to me when a paper of mine was read by a discussant prior to its presentation. I had included in the paper a paragraph discussing Goldstein's concept of the neurobiological aspects of the organism's capacity to transcend its immediate situation, not at all under the impression that I was saying anything very provocative. My using the word *transcending* in introducing the topic, however,

was like waving a red flag in my discussant's face, for he printed a huge "No!!" in red crayon replete with exclamation marks on the margin before even getting to the discussion of what the word meant. The very word, indeed, seems to carry some inciting-to-riot quality.

2. Erwin W. Straus, "Man, a Questioning Being," *UIT Tijdschrift voor Philosophie* 17 (1955).

3. Lawrence Kubie, *Practical and Theoretical Aspects of Psychoanalysis* (New York: International Universities Press, 1950), p. 19.

4. Medard Boss, *Psychoanalyse und Daseinsanalytik* (Bern and Stuttgart: Hans Huber, 1957).

5. Sören Kierkegaard, *The Sickness unto Death,* trans. Walter Lowrie (New York: Doubleday, 1954), p. 163. The quote continues, "Imagination is the reflection of the process of infinitizing, and hence the elder Fichte quite rightly assumed, even in relation to knowledge, that imagination is the origin of the categories. The self is reflection, and imagination is reflection, it is the counterfeit presentment of the self, which is the possibility of the self."

6. Ludwig Binswanger, "The Existential Analysis School of Thought," in *Existence: A New Dimension in Psychology and Psychiatry,* ed. Rollo May, Ernest Angel, and Henri Ellenberger (New York: Basic Books, 1958), p. 197.

7. *Ibid.,* p. 308.

TWELVE Concerning Therapeutic Technique

1. The main exception to this is Irvin Yalom's excellent book *Existential Psychotherapy* (New York: Basic Books, 1980), which is specifically about techniques. But the reader will find in this book not rigid instructions as to what to do in such-and such cases, but rather a discussion of different things a therapist does, or has the possibility of doing, in varied situations.

2. The term *analyzed* itself reflects this problem, and patients may be doing more than using a semantic difficulty as a way of expressing resistance when they aver that the idea of "being analyzed" makes them objects being "worked upon." The term is carried over into the phrase *existential analysis* partly because it has become standard for deep psychotherapy since the advent of psychoanalysis and partly because existential thought itself (following Heidegger) is an "analysis of reality." This term is a reflection of the tendency in our whole culture, called "The Age of Analysis" in the title of a recent survey of modern Western thought. Though I am not happy about the term, I have used the identification "existential analyst" because it is too clumsy to say "phenomenological and existential psychiatrists and psychologists."

3. Quoted by Ulrich Sonnemann in *Existence and Therapy* (New York: Grune & Stratton, 1954), p. 343. Sonnemann's book, we may add, was the first in English to deal directly with existential theory and therapy and contains useful and relevant material. It is therefore the more unfortunate that the book is written in a style which does not communicate.

4. Quoted by Sonnemann, p. 255.

5. C. R. Rogers, "Persons or Science? A Philosophical Question," *American Psychologist* 10 (1955):267–278.

6. This is a point the phenomenologists make consistently—namely, that to know fully *what* we are doing, to feel it, to experience it all through our being, is more important than to know *why.* For, they hold, if we fully know the *what,* the *why* will come along by itself. One sees this demonstrated very frequently in psychotherapy. The patient may have only a vague and intellectual idea of the "cause" of this or that pattern in his behavior, but as he explores and experiences more and more the different aspects and phases of this pattern, the cause may suddenly become real to him not as an abstracted formulation but as one real, integral aspect of the total understanding of what he is doing. This approach also has an important cultural significance. Is not the *why* asked so much in our culture precisely as a way of detaching ourselves, a way of avoiding the more disturbing and anxiety-creating alternative of sticking to the end with the *what?* That is to say, the excessive preoccupation with causality and function that characterizes modern Western society may well serve, much more widely than realized, the need to abstract ourselves from the reality of the given experience. Asking *why* is generally in the service of a need to get power *over* the phenomenon, in line with Bacon's dictum, "knowledge is power" and, specifically, knowledge of nature is power over nature. Asking the question of *what,* on the other hand, is a way of *participating* in the phenomenon.

7. This could well be defined as "existential time"—*the time it takes for something to become real.* It may occur instantaneously, or it may require an hour of talk or some time of silence. In any case, the sense of timing the therapist uses in pondering when to interpret will not be based only on the negative criterion—how much can the patient take? It will involve a positive criterion—has this become real to the patient? As in the example above, has what she is doing in the present to the therapist been sharply and vividly enough experienced so that an exploration of the past will have dynamic reality and thus give the power for change?

8. We are not speaking here of the practical question of what to do when patients actually threaten suicide. This introduces many other elements and is a different question. The conscious awareness we are speaking of is a different thing from the overwhelming and persistent depression, with the self-destructive impulse unbroken by self-conscious awareness, which seems to obtain in actual suicides.

Index

abstract thinking, 51–53, 54, 94, 95, 145
Adler, Alfred, 140
agape, 19, 21
aggression, 61, 82, 106
  acceptance of, 107–8
  neurotic vs. normal, 108
alienation, 65
  loss of world and, 118–22
Allport, Gordon, 47, 80
altruism, 82
Anaximander, 116
Angst, anxiety as, 110–11, 112
animals:
  awareness in, 28
  environment of, 127–28
  loss of centeredness in, 27–28
  sense of time in, 136
anxiety, 16–17, 59, 165, 168
  acceptance of, 107–8
  as *Angst*, 110–11, 112
  anguish vs. dread in, 111
  as birth trauma, 111–12
  contemporary form of, 17
  defined, 109–10, 112
  encounter as source of, 22, 93
  fear vs., 110
  Freud vs. Kierkegaard on, 14–15
  neurotic vs. normal, 108
  as ontological, 33–34, 77, 109–12
  panic and, 78
  psychosis, 32, 33, 109
  as sign of values, 10
  of therapist, 19, 40
  threat of nonbeing and, 105, 107–8, 109–10
  time and, 133, 135, 138, 141
  vigilance as counterpart to, 28
archetypes, 113–14
Aristotle, 34, 137
art, 45, 132
  existentialism in, 48, 56
  fragmentation and, 62
Augenblick, 141–42
Augustine, Saint, 49, 138
awareness, 28–31, 60
  human form of, *see* self-consciousness
  vigilance as, 28, 29

Bacon, Francis, 144
Bally, G., 39
becoming, 50, 57
  being vs., 80–81, 97
  time and, 136, 139
behavioral psychology, 10, 16, 47, 52, 127
being, 9, 10, 91–108
  becoming vs., 80–81, 97
  as "beyond good and evil," 102
  case study of experience of, 98–105
  cultural repression of, 10, 15, 20–21, 95, 105
  ego vs., 103–5
  "I am" experience and, 98–105
  nature vs., 77–78
  Nietzsche's concept of, 80–81
  nonbeing and, 18, 27, 97–98, 105–8
  as potentiality, 17, 18, 97
  problems with definition of, 16, 94–96
  surrender of, to therapist, 10
  as verb form, 50–51, 97
Being and Time (Heidegger), 55, 56
being in the world, 40, 117–25, 129
  loss of, 118–22
  rediscovery of, 117–18, 122–25
Berdyaev, Nicolas, 55
Bergson, Henri, 55, 133, 136–37
Beyond Good and Evil (Nietzsche), 75, 77–78
Binswanger, Ludwig, 38, 39, 41–43, 44, 45, 49, 55, 96, 110, 131, 148, 163
  "*Dasein* choosing" of, 96–97
  "dual mode" of, 93
  "Ellen West" study of, 41–42, 68, 150
  on Freudian psychology, 84–85
  on past time, 139–40
  on therapist-patient relationship, 157–58
  on world, 117, 123–24, 129, 150
biology, biological world, 81, 106
  awareness and, 28–30
  determinism and, 126, 127, 129, 132
  see also animals; *Umwelt*

birth trauma, anxiety as, 111 12
"blind-spots-structuralized-into-dogma," 45
Bohr, Niels, 70
Boss, Medard, 39, 112–15, 116, 127, 154, 155–56
  therapeutic techniques of, 153, 156
  transcendence emphasized by, 148–49
brain-injured, transcendence impaired in, 145, 146
Buber, Martin, 28
Burns, Robert, 146
Buytendijk, F. J., 39

Camus, Albert, 56, 119
caring, 10
  Sorge and, 148–49
Cassirer, Ernest, 64, 86
Castle, The (Kafka), 119
castration, ostracism vs., 21
catharsis, 34
centeredness, 26–28
  loss of, 20, 27–28
  preservation of, 20, 26–27
Cézanne, Paul, 48, 56, 62
character analysis, 44
children:
  irrationality of, 63
  rape fantasy of, 71
  repression in, 16–17
choice, 27, 33, 81, 102, 147, 152
commitment, 13, 69, 128, 152
  in existential therapy, 165–69
  necessity of, 72–73
communication:
  erotic, 21–22
  in existential psychology, 19–20
  technical, 119
  see also language
compartmentalization, 86
  in nineteenth century, 62–66
Concept of Anxiety, The (Kierkegaard), 14–15
Concluding Unscientific Postscript (Kierkegaard), 54
conformism, 16, 20–21, 28, 57, 75, 102
  social response vs., 147
  as threat of nonbeing, 105, 107
conscience, 146
  bad, 82
consciousness, 22, 73, 95, 146, 148
  defined, 29
  tragedy and, 33–34
  see also self-consciousness
control, power vs., 78–79
Copernicus, Nicolaus, 70
courage, 79, 81
  as self-affirmation, 27
Courage to Be, The (Tillich), 55–56
creativity, 116, 167
  encounter and, 22
  sexuality and, 82–83
crisis:
  human experience of, 14, 44, 57
  in modern culture, 56–57, 59, 60–66, 121
culture:
  cure as adjustment to, 164–65
  deviations from Freudianism and, 44
  individual vs., 16
  modern crisis in, 56–57, 59, 60–66, 121
  ontological guilt and, 116
  in repression of being, 10, 15, 20–21, 95, 105
  world vs., 123

Dasein, 25, 91, 101, 128, 148, 164
  defined, 96–97
  Pascal's description of, 57–58
  see also being; existing person
death, 10, 42
  anxiety in relation to, 14
  from suicide, 9, 41–42, 98, 169
  as threat of nonbeing, 14, 33, 105–7, 168–69
death instinct, 33, 106
decision, 13, 27, 128, 166–67
Dehumanization of Art, and Other Writings on Art and Culture, The (Ortega y Gasset), 57
depression, 60–61
  time and, 133, 135–36, 138
Descartes, René, 120, 121
despair, 9, 10, 53–54, 57, 60–61, 65
  individual vs. mass, 78
determinism, 25, 85, 96
  biological, 126, 127, 129, 132
  social, 127
  time and, 139, 141
dignity of man, 76–77, 81
Dilthey, Wilhelm, 55
Doll's House, A (Ibsen), 62
Dostoevski, Fëdor, 34, 48
dreams, 31–33, 69, 112–14, 127
  decision reflected in, 166–67
  "I am" experience and, 98–105
  in psychoanalysis vs. existential therapy, 152

Eastern languages, for-me-ness in, 129
Eastern philosophies, existentialism as similar to, 58–59
ego, 14, 18, 44
  "I am" experience vs., 103–5
  passivity of, 60, 74, 85, 103–5

Eigenwelt, 85, 102, 123, 128–32, 146
  guilt corresponding to, 115
  literal meaning of, 21*n,* 126
  in therapeutic process, 162–63
  time in, 137
  transcendence and, 147–48
Ellenberger, Henri, 46
"Ellen West," Binswanger's study of, 41–42, 68, 150
emotion:
  in therapeutic relationship, 21–22, 23, 72
  Victorian views on, 62–63
encounter, concept of, 19–23
  anxiety vs. joy in, 22
  change in, 22, 128
  as creative experience, 22
  instantaneous, 91–92
  in therapy, 19, 21–23, 91–94
  transference as distortion of, 19, 23
Enlightenment, ecstatic reason in, 85, 86
environment, *see Umwelt*
epiphany, 142
eros, 19
  therapeutic relationship and, 21–22
essence, existence vs., 51
esteem, 21, 102, 110
ethics, 16, 85
evolution, emergent, 143–44
existence:
  emergence of, 136, 143–44
  essence vs., 51
  etymology of, 50
  experiencing the reality of, 162–65
Existence, 44
existential-analytic movement, 38–48
existentialism, 48–59
  birth of, 54
  confusion about, 48–49
  as cultural movement, 48, 50
  cultural situation giving rise to, 60–66
  defined, 49
  Oriental thought and, 58–59
  pessimism vs. optimism of, 57
  phases in development of, 54–55
  psychoanalysis compared to, 59
existential psychology:
  criticism of and resistance to, 13, 16, 44–48, 94–95
  as dynamic approach, 50–51, 69–73
  origins and significance of, 37–59
  previous deviations from Freudianism vs., 43–44
  as protest against technical reason, 87–88
  repression in, 16–18, 154–55
  structure of human existence as emphasis of, 43, 44, 159–60
  tragedy and, 33–34
  transference in, 16–23, 154, 160–61
  see also therapeutic technique
existing person, 91
  six principles of, 25–34
  use of term, 25
  see also Dasein
experience:
  denial of, 94–95
  "I am," 98–105
  quality of, 91–92

fantasy, 21, 30, 71
fear, anxiety vs., 110
Feuerbach, Ludwig, 50
fragmentation, 56
  of nineteenth-century personality, 60, 61–66, 105
  therapy as expression of, 9–10, 86–87, 164, 165
freedom, 18, 29, 82, 145, 155, 165
  burden of, 34, 112
  as need, 10
  in social behavior, 147
  transcendence as basis of, 149–50
  truth as, 70
Freud, Anna, 43
Freud, Sigmund, 16, 30, 39, 41, 60, 74, 96, 105, 129, 131, 139, 171
  anxiety as viewed by, 14–15, 110, 112
  Binswanger's relationship with, 42–43
  Kierkegaard compared to, 68–69, 84
  Nietzsche as viewed by, 61, 83
  Nietzsche compared to, 68–69, 82, 83–84
  technical emphasis of, 14–15, 65, 84–87, 106
  translucence of, 156
  see also psychoanalysis
Fromm, Erich, 44, 119, 158
Fromm-Reichman, Frieda, 23, 118, 158
future, 97, 123, 135, 138, 167
  existential analysts' emphasis on, 139–40
  time binding and, 136

Genealogy of Morals (Nietzsche), 82, 83
Gestalt, 29, 38, 130, 142, 167
"God Is Dead" (Nietzsche), 76
Goldstein, Kurt, 29, 109, 110, 112, 144, 146
Groddeck, Georg, 103
Guernica, 56
guilt, 53–54
  against fellow humans, 115
  from forfeit of potentialities, 112–15
  guilt feelings vs., 113, 114
  neurotic, 116
  as ontological, 112–16
  separation, 115–16

harmony, doctrine of pre-established, 121
health, Nietzsche's views on, 80, 81
Hegel, Georg W. F., 49, 54, 69
Heidegger, Martin, 38, 55, 56, 70, 77, 105, 148, 149
Heisenberg, Werner, 70
historical progress, psychotherapy and, 140–41
Horney, Karen, 44, 106
hostility, 16, 53–54, 61, 63, 82, 93, 106
  acceptance of, 107–8
  neurotic vs. normal, 108
human:
  absence of unified view of, 64
  as machine, 15, 50, 54, 63–64, 69, 87
  new conception of, 38
human being, meaning of term, 97
humility, guilt as basis of, 115, 116
Husserl, Edmund, 55, 144
Hutchens case, 24–34
  characteristics of patient as existing person in, 25–34

"I am" experience, 98–105
  ego vs., 103–5
  as precondition to solution of problems, 100–101
Ibsen, Henrik, 62
id, 74, 103, 105
individual:
  existing person vs., 25
  of Kierkegaard, 69, 77
industrialism, compartmentalization and, 63–64
inference, knowledge of world by, 120–21
inhibition, 28
  defined, 20
inner certainty:
  being as source of, 10
  objective truth vs., 9
insight, 31, 130, 166
  time and, 137–38, 141–42
instinct, repression of, 60, 61, 82
intellectualism, 50, 53, 129
interest, use of term, 69
interpersonal psychology, world concept and, 130, 131
Interpretation of Dreams (Freud), 69
interviews, feelings in, 92–93
irrationality, 74, 85
  Victorian views on, 63
  see also unconscious
isolation, 94
  loss of world and, 118–22

James, William, 13, 55, 78
Jaspers, Karl, 15, 16, 53, 55, 157, 158
Jones, Ernest, 83, 84
joy, 133
  encounter as source of, 22, 93
  power as, 80, 83
Jung, C. G., 22
Jungian analysis, interpretation of dreams in, 112–14

Kafka, Franz, 48, 56, 119
Kairos (time fulfilled), 142
Kaufmann, Walter A., 79–80
Kierkegaard, Sören, 17, 30, 52–53, 54, 65, 67–73, 127, 166
  analysis of time and, 139, 141–42
  on anxiety, 14–15, 53, 60, 67, 110, 112
  Freud compared to, 68–69, 84
  on imagination, 149
  on necessity of commitment, 72–73
  Nietzsche compared to, 52–53, 65–66, 68–69, 73–75, 76–77, 83–84
  rationalism protested by, 49, 62, 85–86
  on relational truth, 69–72
  on self-estrangement, 60–61, 118
know, to, etymology of, 93
knowing vs. knowing about, 15, 92–93
knowledge:
  of death, 106
  decisions in relation to, 166
Kubie, Lawrence, 148
Kuhn, Roland, 39, 96, 124–25

Laing, R. J., 154
language:
  evidence of estrangement from nature in, 119, 122
  for-me-ness in, 129
  of "outer-directed" persons, 119
Laotzu, 58
Lefebre, Ludwig, 40
Leibnitz, Gottfried von, 47, 121
libido, 14, 16–17, 40, 103, 106, 129–30
Liddell, Howard, 28, 29, 127, 136
literature, 45, 142
  existentialism in, 48, 56
  fragmentation in, 62
loneliness, 21, 65, 94
  epistemological, 119
  loss of world and, 118–19
Lonely Crowd, The (Riesman), 119
love, 10, 21–22, 95, 141
  as agape, 19, 21
  Freud's views on, 18, 19
  knowing and, 93
  transference in, 20
  world concept and, 128, 131–32, 137
lying, transcendence and, 146

Macquarrie, John, 55
Marcel, Gabriel, 55, 95–96
Maritain, Jacques, 49
Marx, Karl, 54–55, 63–64, 119
Marxism, 105, 141
Maslow, Abraham, 80
materialism, existentialism as, 48, 49
mathematics, 94, 120
meaning:
  borrowed, 21
  in existential therapy, 154–56
Meaning of Anxiety, The (May), 14
Middle Ages, 120
Minkowski, Eugene, 39, 133–35, 138
Mitwelt, 21, 23, 85, 127–28, 132, 146, 155
  guilt corresponding to, 115
  in interpersonal theory, 130, 131
  literal meaning of, 21*n,* 126
  in psychoanalysis, 129–30
  time in, 137
monads, reality and, 121
morality, 82, 102, 154–55
moths, gypsy, loss of centeredness in, 27–28
Mowrer, O. Hobart, 136
mystic, as derogatory term, 94–95

nature, 81
  alienation and isolation from, 58, 119–22
  being vs., 77–78
  Copernican view of, 70
  separation guilt and, 115–16
  see also Umwelt
neurology, relations to real world and, 120–21
neurosis, 16–17, 23, 28, 30, 60, 162, 164
  as adjustment to preserve centeredness, 26–27
  contemporary pattern of, 17, 20–21, 28
  guilt as, 116
  individual vs. society and, 65
  structuralizing of, 152
  as symbolic process, 148
New York Times, 48–49
Nietzsche, Friedrich, 13, 17, 52–53, 55, 73–84, 102, 139, 143, 146, 169
  being concept of, 80–81
  on effects of industrialism, 63–64
  Freud compared to, 68–69, 82, 83–84
  "impotent people" of, 82, 108
  Kierkegaard compared to, 52–53, 65–66, 68–69, 73–75, 76–77, 83–84
  ontological meaning in, 77–78
  prophetic views of, 9, 61, 65 66, 118
  "will to power" of, 75–78
nonbeing, 18, 33, 97–98, 105–8, 168–69
  acceptance of, 27
  as cure, 164–65
  possibility of annihilation and, 9, 10
  resistance to use of term, 16

objectivity, 69, 70
  see also subject-object cleavage
obsessional-compulsive, case study of, 112–15
ontological principles of the existing person, 25–34
ontology, 15, 91
  defined, 51
  meanings of psychological terms and, 77–78
  see also being
Organization Man (Whyte), 15–16
Ortega y Gasset, José, 55, 56–57
ostracism, fear of, 21

participant observation, 72
participation, 27–28, 93
  in ontological guilt, 116
  risk of, 20, 27–28
  self-dispersal in, 20–21
Pascal, Blaise, 49, 57–58, 98
patients:
  pre-appointment imaginings of, 20
  presence and, 156–61
  world of, 38, 117–18, 122–25
phenomenology, 23, 39, 55
Phenomenon of Man, The (Teilhard de Chardin), 29
Philosophical Fragments (Kierkegaard), 54
philosophy, 45
  assumptions of science and, 46, 51
  Eastern, 58–59
  encroachment into psychiatry of, 45–46
  Freud's views on, 83–84
  see also existentialism
physics, reversal of Copernican view of nature in, 70
Picasso, Pablo, 48, 56
pleasure, unimportance of, 17, 80
Pope, Alexander, 159
power, 17, 21, 81
  control vs., 78–79
  experience of being vs., 100–101
  as *potentia,* 79–80, 82
  technical, 9
  "will to," 75–78
pragmatism, 47, 55
presence, in existential therapy, 156–61
Problem of Anxiety, The (Freud), 14–15
promises, as characteristic of humans, 146
Protestant Era, The (Tillich), 121
psychoanalysis:
  attempts at systematization of, 25
  basic rule of, 156

psychoanalysis *(continued)*
concept of man in, 40
cultural situation giving rise to, 60–66
death instinct of, 33, 106
deviations from, 43–44
ego concept in, 103–5
existentialism compared to, 59
interpretation of dreams in, 112–14, 152
Mitwelt in, 129–30
nature of cure in, 164–65
past time emphasized in, 140, 153–54
transference in, 18, 20, 23, 154
see also Freud, Sigmund
psychological terms, ontological meaning of, 77–78
psychology:
behavioral, 10, 16, 47, 52, 127
depth, 53–54, 68, 74, 84, 128
interpersonal, 130, 131
Lockean vs. Leibnitzian tradition of, 47
technical vs. existential level in, 14–15, 25, 26, 39–40, 47, 50, 51–53, 86–88, 91–94, 151
see also existential psychology
psychosis anxiety, 32, 33, 109
psychotherapy:
bases of, 13–23
data collection vs. encounter in, 91–94
existential psychology's contributions to, 91–171
as expression of fragmentary age, 9–10, 86–87, 164, 165
levels of patient-therapist relationship in, 21–22
see also therapeutic technique

Rank, Otto, 43–44, 112
rape, childhood, 71
rationalism, 49, 54, 62–63, 74
re- (prefix), 146–47
reaction formation, 82
reality, 58–59
as lawful, 46
mathematics as basis of, 94
monads and, 121
technical view as distortion of, 93–94
truth vs., 49, 51–53
reason, 62, 145–46
ecstatic, 85–86
technical, 64, 85–87, 106
Reich, Wilhelm, 44
relationship, 19–20
in *Mitwelt,* 127–28, 130
presence and, 156–61
to time, 134–35, 138
truth as, 69–72, 156–57
see also encounter, concept of; participation
relationship therapy, 102
religion, 31–32, 68, 69, 121, 141, 142
dreams and, 113–14
fragmentation and, 62
Nietzsche and, 75, 76
repression, 148
of being, 10, 15, 20–21, 95, 105
in existential psychology, 16–18, 154–55
in Freudian psychology, 16, 30, 60, 82, 153–54
inhibition and, 20
struggle of being vs. nonbeing in, 18
time disturbances and, 138
in Victorian period, 64–65
resistance, 154, 155–56
responsibility, 15, 18, 146–47, 148
transference and, 19
Riesman, David, 119
Rilke, Rainer Maria, 48, 124
risk, 22
of participation, 20, 27–28
Robinson, Edward, 55
Rogers, Carl, 158–59, 161
"Rudolf R.," case study of, 124–25

Sargent, Helen, 46
Sartre, Jean-Paul, 48–49, 55, 77, 81, 94, 146
Schachtel, Ernest, 63, 123
Scheler, Max, 64, 86
Schelling, Friedrich von, 54
schizophrenia, 68, 118
Schopenhauer, Arthur, 54, 82–83
sciences:
autonomous, 64
essentialist metaphysics of, 51
scientific method, 45–46
security, 16–17, 22
self-actualization, "will to power" as, 79
self-affirmation, 10, 27, 108
self-consciousness, 28–33, 56, 57, 67, 129
in "I am" experience, 101–2
of Kierkegaard and Nietzsche, 74–75
loss of, 15–16
ontological guilt and, 116
transcendence and, 143–44, 147–48
self-esteem, 102, 110
self-estrangement, Kierkegaard on, 60–61
sexual intercourse, 93
Freudian view of, 18
sexuality, 71
artistic energy and, 82–83
repression of, 16, 63, 75, 153–54
self-consciousness and, 30
sexual objects, 30, 128
Shakespeare, William, 135–36

"Sickness unto Death" (Kierkegaard), 42, 68
social relationships:
  role of transcendence in, 146–47
  see also Mitwelt
social sciences, conformism encouraged by, 16
Socrates, 49, 57, 161
Sorge (care), 148–49
space, time vs., 96, 133, 136–37
spatial relations, 122
Spence, Kenneth W., 52
Spinoza, Baruch, 85, 86
Storch, A., 39
Stranger, The (Camus), 56, 119
Straus, Erwin, 39, 40, 117, 120–21, 137, 147
subject-object cleavage, 129, 167
  being in the world and, 118, 120–22
  ego and, 104
  existentialism as attempt at overcoming of, 49–50, 53, 55, 58–59, 70–71
suicide, 9, 41–42, 98, 169
Sullivan, Harry Stack, 28, 44, 72, 74, 118
  interpersonal theory of, 130, 131
superego, 102, 103
survival value, 17, 82
Suzuki, 129
symbols, use of:
  in dreams, 113–14
  in neurosis, 148
  transcendence and, 145–46, 148
symptom formation, 82, 116
Szasz, Thomas, 19

technical reason, 64, 85–87, 106
technology, negative effects of, 9, 15
Teilhard de Chardin, Pierre, 28, 29
tenderness, 17, 19
therapeutic technique, 151–71
  commitment and, 165–69
  context of, 152
  danger of generality in, 169–70
  experiencing the reality of existence as goal of, 162–65
  meaning and, 154–56
  presence and, 156–62
  timing and, 160–61
  understanding in relation to, 151–52
  variability of, 153–54
therapists:
  as agents of conformism, 15–16
  defenses of, 19, 160
  doubts of, 37–38, 39–40
  feelings of, 21–22, 23, 72
  participant observation of, 72
  presence and, 156–61
Thompson, Clara, 21
thrownness, 139–40
  use of term, 58

Tillich, Paul, 33, 55–56, 79, 85, 108, 119, 121, 142
  on courage to be, 27, 81
time, 58, 133–42
  binding of, 136
  clock, 137
  present, Rank's emphasis on, 43–44
  space vs., 96, 133, 136–37
  transcendence and, 144, 145
  see also future
tragedy, 106, 132
  confronting of, 33–34
transcendence, 19, 21, 143–50
  consciousness and, 29
  defined, 143
  neurobiological base of, 145
  problems with term, 144–45
  time and, 136
transference, 40
  as defense for therapist, 19, 160
  as distortion of encounter, 19, 23
  in existential psychology, 16–23, 154, 160–61
  in Freudian psychology, 18, 20, 23, 154
  "I am" experience and, 101–2
  in love, 20
trust, 20, 21, 101, 146
truth, 69–73
  commitment to, 72–73
  as inwardness or freedom, 70
  Nietzsche's views on, 75–76
  as produced in action, 55, 72–73
  reality vs., 49, 51–53
  as relationship, 69–72, 156–57

Uexküll, Jakob von, 122, 149–50
Umwelt, 41, 85, 122, 126–27, 129, 130, 132, 146, 162
  guilt corresponding to, 115
  literal meaning of, 21*n*, 126
  love and, 131
  time and, 137, 139
Unamuno, Miguel de, 55
unconscious, 40, 74, 86, 148
  commitment and, 167
  decisions and, 166
  in existential psychology, 17–18, 170–71
  in Freudian psychology, 16

values, 17, 81, 121
  anxiety and, 10
  as need, 10
Van Den Berg, J. H., 39, 154
Van Gogh, Vincent, 48, 56, 62
Victorian period, compartmentalization and inner breakdown in, 62–66, 75, 105

Vienna Psychoanalytic Society, 39, 83
vigilance, as awareness of threats, 28, 29
von Gebsattel, V. E., 39

Way of Life, The (Laotzu), 58
Whitehead, Alfred North, 55
Whyte, William, 15–16
Wild, John, 51
will, 27
  ontological meaning of, 77–78
  Victorian views on, 62
"will to power," 75–78
wishes, will in relation to, 27
world, 103–5, 117–32
  being in the, 40, 117–25, 129
  culture vs., 123
  as dynamic pattern, 123–24
  inference and, 120–21
  loss of, 118–22
  openness of, 123
  of patient, 38, 117–18, 122–25
  rediscovery of, 117–18, 122–25
  as spatial relation, 122
  as structure of meaningful relations, 122–23
  three modes of, 126–32; *see also* *Eigenwelt; Mitwelt; Umwelt*
World as Will and Idea, The (Schopenhauer), 54

Zen Buddhism, 58